Statistical Concepts

Statistical Concepts
A Basic Program

Foster Lloyd Brown
State University of New York, Oneonta

Jimmy R. Amos
Eastern Idaho Community Mental Health Center

Oscar G. Mink
University of Texas

SECOND EDITION

HarperCollins*Publishers*

Sponsoring Editor: George A. Middendorf
Project Editor: Karla B. Philip
Designer: Andrea Clark
Production Supervisor: Will Jomarrón

STATISTICAL CONCEPTS: A Basic Program, Second Edition

Library of Congress Cataloging in Publication Data

Brown, Foster Lloyd.
　　Statistical concepts.

　　On the 1st ed. Amos' name appeared first on t. p.
　　Bibliography: p.
　　1. Mathematical statistics—Programmed instruction.
I. Amos, Jimmy Ray, joint author.　　II. Mink, Oscar G.,
joint author.　　III. Title.
QA276.A59　　1975　　　　519.5′077　　　74-17851
ISBN 0-06-040988-6

Contents

Preface to the Second Edition

The first edition of this book has been one of the most successful books of its kind on the market. Consequently, this second edition has been carefully restructured to expand opportunities for the reader without sacrificing the brevity and clarity of the first edition.

The second edition differs from the first principally in the addition of (1) practical calculating formulas throughout the text, (2) mastery quizzes at the end of each section, and (3) the significant expansion of the sections on analysis of variance, regression, confidence intervals, and chi-square. Additional changes include the condensing of the sections on measurement (at the end of the book) and a clearer segmentation of the program into twelve modules.

It is hoped that this second edition will continue to be useful as a course supplement and that, with the addition of suitable exercises, it will also serve as a core text for quarter- and one-semester courses in statistical methods.

I am indebted to the Literary Executor of the late Sir Ronald A. Fisher, F.R.S., to Dr. Frank Yates, F.R.S., and to Longman Group Ltd., London, for permission to reprint Tables III, IV, V, and VI from their book *Statistical Tables for Biological, Agricultural and Medical Research*.

Foster Lloyd Brown

Preface

The beginning student in psychology, educational psychology, guidance, and similar fields in the behavioral sciences must gain a conceptual understanding of statistical procedures in order to read the experimental literature assigned for readings in most introductory courses.

This program was developed in an effort to assist our students in obtaining a conceptual understanding of certain statistical phenomena without sacrificing a week of class time presenting the material in lecture and discussion sections. Our reasons went beyond the desire to conserve class time for other introductory material however, as our experience with a large class of students with diverse abilities suggested that some were quick to grasp or already knew the basic statistical concepts. As a result, some students were easily bored, whereas others needed special review sessions. We believed that the ideal answer to our dilemma was completely individualized instruction. Our solution was programmed instruction. We searched for a usable program. To our dismay we could locate nothing that seemed appropriate; the programs were either too comprehensive—or too much like workbooks and programs in name only.

This program exposes the student to: (1) Descriptive statistics—central tendency, measures of variability, correlation, and the application of these concepts in the concepts of reliability, validity, and norming; (2) Inferential statistics—significance testing including analysis of variance.

The program can be completed in five hours by the typical beginning student. The program has been tested seven times for error rate and has been used in seven separate experimental studies. The evaluative data available suggests that the program is fulfilling its purpose. As you use this program, you will find that it contributes to your effectiveness as instructors and students. Many of you will have creative suggestions and constructive criticisms of value to the authors. We welcome such assistance and will embody the best suggestions in future editions.

Jimmy R. Amos
Foster Lloyd Brown
Oscar G. Mink

Tips to the Student

There are no doubt some of you who are somewhat reluctant to study statistics because you have heard that it is difficult and highly technical. This text was carefully designed to present the material in small steps and to check your growing understanding at each step. Even the most mathematically inept should be able to understand and enjoy this book.

The book is presented in a series of numbered *frames*. Each frame consists of three parts: (1) the explanation of a statistical concept, (2) a question based on the concept, and (3) the corresponding "book response" to the question.

Try to respond correctly to each frame before you see the "book response." It will help you to avoid seeing the book response if you use a mask of paper or cardboard which you slide down the page as you progress. You can thus make sure that you understand each frame as it comes along. This is especially helpful when you are tired or preoccupied. A programmed text is well adapted to study in odd free moments that might otherwise be wasted. If you are a particularly busy person, you might wish to take advantage of this characteristic. The very best results will occur if you can give this text the same undivided attention you would give a classroom text.

Some of you will spend less than five and one-half hours working the program; but, you will have mastered some important concepts necessary for understanding today's research literature in the behavioral and social sciences.

Frequency Distribution

1. Such things as test scores, class rank, weight, and income are called *variables*. Income, for instance, is called a variable because different income values are possible. In general, things that vary in value from case to case or time to time are called _____.

   Variables

2. The number of times a particular value of a variable occurs is referred to as the *frequency* of that value. If 17 students receive a score of 70 on a test, then the score of 70 has a _____ of 17.

   Frequency

3. A distribution is a series of separate values such as scores which are arranged or ordered according to magnitude. A group of ordered scores is a *distribution*. For example, a group of scores ranging from the lowest to the highest score is a _____ (see table).

Scores
13
11
11
9
9
9
8
5

   Distribution

4. A set of ordered scores and their corresponding frequencies is called a *frequency distribution*. This can be represented in

table or graph form. The table below shows the number of times a score occurs in its group. This table is a frequency

_____ .

Scores	Frequency
13	I
11	II
9	III
8	I
5	I

Distribution

5. Frequency distributions can also be graphically illustrated. The two most common graphs used to illustrate frequency distributions are the *frequency polygon* and the *histogram*. If scores and their frequencies are illustrated with points connected by lines, it is called a *frequency polygon*. Because the illustration below shows the frequency of particular scores by the height of points that are connected by lines, it is called a frequency

_____ .

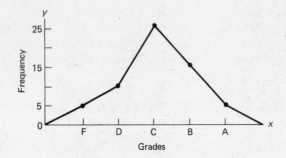

Polygon

6. When a frequency distribution is illustrated in the form of a histogram, the scores and their frequencies are designated by rectangular boxes. In the frequency distribution below, the height of the rectangular boxes indicates the frequency

with which students received particular scores. It is called a
_____.

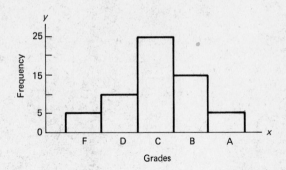

Histogram

7. It is the accepted practice for the vertical side of a graph, called the *ordinate* axis, to be used to designate the frequency. The horizontal side, called the *abscissa* axis, is used for the scores. Direction of increase is upward for the frequency on the ordinate axis. Direction of increase for the variable is from left to right on the_____ axis.

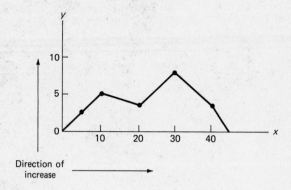

Abscissa

3

8. On this graph the *f*, which designates the frequency, is the
_____ axis, and the *x*, designating the variable,
is the _____ axis.

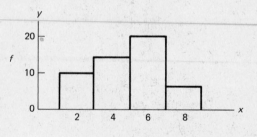

9. The two most common graphs used to illustrate frequency distributions are the frequency polygon and the _____.
Graph A is a _____. Graph B is a _____
_____.

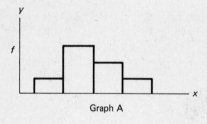

Graph A

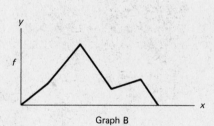

Graph B

1. Construct a frequency distribution to represent the following scores: 0, 2, 1, 2, 3, 2, 2, 3, 3, 5.

2. Draw a histogram to illustrate the frequency distribution in Problem 1.

3. Using a dotted line, draw a frequency polygon to represent the frequency distribution in Problem 1. Superimpose this drawing over the histogram drawn in Problem 2.

4. For the following frequency polygon, label the abscissa and the ordinate.

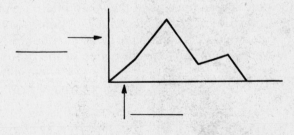

5. Frequency is designated on the _____ axis.

1.

Scores	f
0	I
1	I
2	IIII
3	III
4	0
5	I

See Frame 4

2. and 3. See Frames 5 and 6

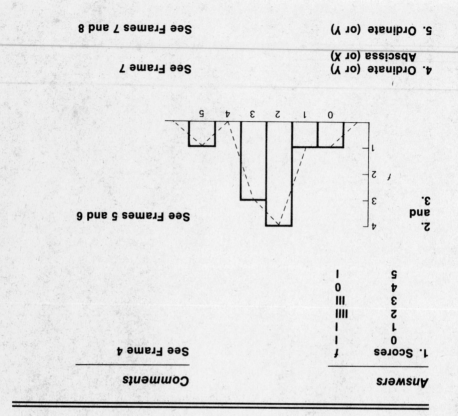

4. Ordinate (or Y)
   Abscissa (or X)   See Frame 7

5. Ordinate (or Y)   See Frames 7 and 8

Averages

10. After scores have been tabulated into a frequency distribution, a measure of *central tendency*, or central position is often calculated. Central tendency gives us a concise description of the average or typical performance of the group as a whole. Measures of _____ tendency allow us to compare two or more groups in terms of typical performance.

Central

11. In statistics there are several "averages" or measures of

_____ _____ in common use. Three of these are: (a) the mean, (b) the median, and (c) the mode.

Central Tendency

12. The mean is generally the most familiar and most useful to us. The mean is computed by dividing the *sum of the scores* by the *total number of scores*. The formula for the mean would be

$$\text{Mean} = \frac{\text{sum of the scores}}{?}.$$

Total Number of Scores

The Mean: Practical Computation

This section on the mean is not so much to teach you how to compute a mean (which you can probably already do), but to introduce some symbols and a way of learning how to do more complex computations.

Using the notes and example problem, solve the practice problem on the right.

Notes	Example problem scores (X)		Practice problem (X)	
13. Sum up the scores and label the sum as ΣX. Σ (the capital Greek letter sigma) may be read as "sum of." X stands for scores.	Don	0	Ima	0
	Ray	4	Lil	2
	Jan	2	Dot	2
	May	4	Hal	3
	Joy	6	Sue	8
	Jim	4		
	Sam	4		
	Fay	6		
	Art	6	___ = ___	
	ΣX = 36			
			ΣX	15

14. *N* symbolizes the number of scores.

$N = 9$ $N = $ _____

5

15. The mean, symbolized by M (or $\bar{X}$, read "X bar"), is the ΣX divided by N.

$$M = \Sigma X/N$$
$$= 36/9$$
$$= 4$$

$M = \underline{\hspace{2cm}}$

$$\overline{\overline{3}}$$

16. Finding the arithmetic mean of a distribution is analogous to finding the center of moment, or the balance point, in a solid block. If a distribution were suspended by the mean, it would hang level or balanced. The mean, whose symbol is _____, is the center of moment in a frequency distribution.

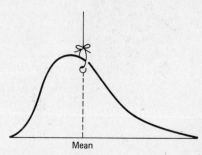

Mean

$$\overline{\overline{\bar{X} \text{ or } M}}$$

17. Thus, if extremely high or extremely low scores are added to a distribution, the mean tends to shift toward those scores. If the center of gravity of the distribution is shifted to one side or the other of the curve, the curve becomes "skewed." The following curve has a few extremely low scores. Consequently, this distribution is _____.

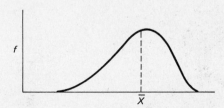

$$\overline{\overline{\text{Skewed}}}$$

8

18. Extreme scores, either high or low, tend to _____ a distribution.

19. If a distribution is massed so that the greatest number of scores is at the right end of the curve and a few scores are scattered at the left end, the curve is said to be *negatively* skewed. If the massing of scores is at the left end of the curve with the tail extending to the right end, then the curve is *positively* skewed. Graph A illustrates _____ skewness. Graph B illustrates _____ skewness.

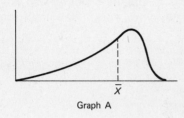

Graph A

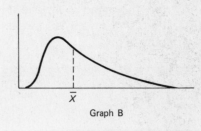

Graph B

20. This graph's tail is extending to the right because of a few extremely high scores and is therefore _____ skewed.

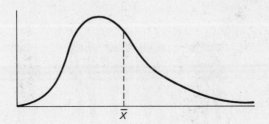

21. A curve is *symmetrical* when one half of the curve is a mirror image of the other half. If you folded a frequency polygon at the mean and the two halves were similar, then the frequency distribution represented by the polygon would be said to be _____.

9

22. According to the formula for computing the mean ($\bar{X} = \Sigma X/N$), we can define the mean as the arithmetic average of the scores in a distribution. If we added extreme scores to one end of a previously symmetrical curve, the mean would shift toward those extreme scores. Would the curve be symmetrical or not symmetrical? _____

Not Symmetrical (or Asymmetrical)

23. Regardless of whether the curve is symmetrical or asymmetrical, the mean is always the center of balance. Does this imply that the mean is always centrally located in asymmetrical curves? _____

No

24. Let us illustrate this point by placing a distribution along an interval scale such as that below. Each figure represents one person. The scale would obviously balance if a fulcrum were under the middle number, 4. To verify this, calculate the mean by the formula $\bar{X} = \Sigma X/N$. Was this distribution symmetrical?

_____

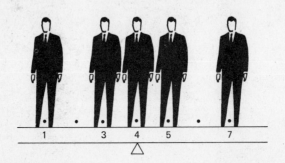

Yes   **($\bar{X} = 20/5 = 4$)**

25. If the person with a score of 7 had scored 12, what would be the mean? _____ Place a fulcrum (i.e., Δ) at

10

the balance point of the scale below. Is the fulcrum centrally located? _____ Is this distribution symmetrical? _____

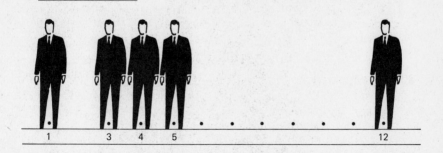

1   3   4   5   12

_____

5   Fulcrum Should Be Under Number 5   No   No

26. What would be the mean for the above distribution if the person who scored 12 had instead scored 22? _____

$$= \frac{}{7}$$

27. When a curve is positively skewed (see Graph A) the mean is located to the _____ (right or left) of most of the cases. When a curve is negatively skewed (see Graph B) the mean is located to the _____ of most of the cases. (Each dot is one case.)

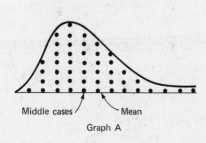

Middle cases   Mean

Graph A

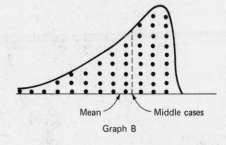

Mean   Middle cases

Graph B

_____

Right   Left

28. The distribution of scores: 2, 2, 3, 3, 15 would be _____ skewed.

_____

Positively

11

29. The preceding distribution of 2, 2, 3, 3, 15 has a mean of 5. What will the mean be if 10 points are added to the score of 15 (making it a score of 25)? _____

$$= \over 7$$

30. We saw that by adding 10 points to the score of 15, the mean of the distribution 2, 2, 3, 3, 15 was raised by 2 points. The reason for this is that the mean is an arithmetic average and each score contributes to its value. When 10 was added it was averaged or distributed equally among the five scores. This has the same effect as adding a constant of 2 points to each score (10 points/5 scores = 2 points per score). When 2 points are added to each score, the mean is raised by _____ points (from 5 to 7).

$$= \over 2$$

31. When a constant is added to each score of a distribution, that constant is added to the previous mean to find the new mean. If each score of a distribution is *multiplied* by a constant, the new mean is found by multiplying the old mean by that _____.

$$\over \text{Constant}$$

32. The distribution 0, 2, 2, 3, 13 has a mean of 4. What would the mean be if each score was *multiplied* by a constant of 2? _____.

$$= \over 8$$

Median

33. By adding or substituting an extreme score to a distribution, the mean no longer represents a centrally located score but represents a measure that is more typical of the extreme score. This causes us to rely on another measure of central tendency

12

which is called the median or the middle score. The median is abbreviated Md or Mdn. The measure of central tendency that is less affected by the addition of an extreme score is the _____.

Median

34. The median is a *point* on a scale of measurement above which are exactly half the cases and below which are the other half of the cases. The student should note that the median is defined as a *point* and not as a specific measurement, for example, a score or a case. From the distribution 4, 6, 8, 10, 12, it is easy to see that 8 is the middle score. The score of 8 is at the *point* where there are two scores above and two scores below, hence, 8 is the median. What is the median of 11, 11, 14, 19, 19? _____

14

35. To obtain the median, the measures are arranged in ascending order from the lowest to the highest measure. Then by counting up this scale, the point is selected above and below which there are an equal number of cases. The value of this point is the middle or the _____ case.

Median

36. The median of the distribution 3, 2, 0, 1, 6 can be found by first arranging the measures from the lowest to the _____ number (0, 1, 2, 3, 6). Then we find the middle score or case, which is _____, and that is the median.

Highest   2

37. When there are an even number of cases, simply average the two middle scores. The median of 0, 6, 9, and 12 is _____.

7.5

38. When a distribution has a midpoint score with a frequency greater than 1 (e.g., 5, 6, 9, 9, 9, 11), use the midpoint score as the median just as before (median is thus 9 for this case). There is a more exact solution but it is seldom used and the exact answer, when rounded to the nearest whole number, is *always* the same as that obtained by the method given here. What is the median of the distribution: 0, 2, 4, 4, 4, 4, 6, 6, 6? _____

$$= \\ 4$$

39. The value of the median as an indicator of central tendency is increased when the end score is extreme. For example, the median of 0, 6, 9, 10, 10 is _____ . The mean of this distribution is 7 (i.e., 35/5). If the extreme number 55 is substituted for one of the 10's, giving the distribution: 0, 6, 9, 10, 55, the median remains _____ but the mean is now _____ .

| 9 | 9 | 16 |

40. When we want to minimize the effect of one or more extreme scores, we should use the _____ to represent the average score of the distribution.

Median

41. The median, for both odd and even number of cases, is the point on a distribution where there are an equal number of cases above and _____ that point.

Below

Mode

42. A third measure of central tendency is the mode. It may be defined as the one value or score that occurs with the most frequency. The mode of the series 2, 3, 4, 4, 4, 5, 5 is 4. The mode of the series 7, 8, 10, 10, 10, 11, 11 is _____ . The median is _____ .

| 10 | 10 |

43. Is it possible for a distribution to have a median and a mode of the same value? _____

44. The mode is used as a simple, inspectional "average" to show quickly, the center of concentration of a frequency distribution. What is the mode or the rough average of the frequency distribution shown below? _____

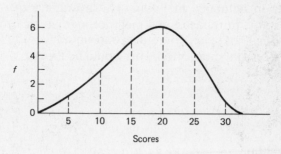

Scores

45. The mode is not generally used unless there is a large number of cases in a distribution. When the number of cases in a distribution is small, it is more likely that several scores will have the same frequency. The frequency polygon shown below is an extreme example. It is evident that the mode is 10 but it does not give a close examination of the average case. The mean is 25 ($\Sigma X/N = 125/5$). The cases, in ascending order, are 10, 10, 25, 35, 45, with the number 25 at the midpoint; thus _____ is the median.

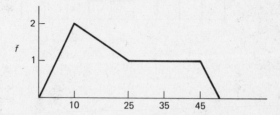

46. The mode is used, in preference to either the median or the mean, when a measure of the most characteristic value of a group is desired. What is meant by "the most characteristic value" can be exemplified by clothing fashions. The _____ is what is being worn the most.

Mode

47. The mode is used also to be sure that the average you obtain exists in actuality. In finding the average size of automobile tire that is purchased, the mean or median size might be a tire that does not exist. Therefore, one would want to know the size of tire bought most often. This would be the _____ .

Mode

48. In addition to serving as a measure of central tendency, the concept of modality is useful in describing the shape of some distributions. If a histogram or a frequency distribution has two peaks, it is referred to as a *bimodal* distribution. If a distribution has more than two peaks, it is called *multimodal*. The following histogram appears to have two separate concentrations of frequencies; consequently it can be described as _____ .

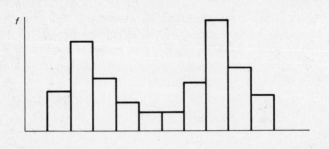

Bimodal

49. The shape of the frequency distribution illustrated below by the frequency polygon is _____ . The distribution of the histogram is _____ . The fre-

16

quency polygon is negatively skewed, whereas the histogram is _____ skewed.

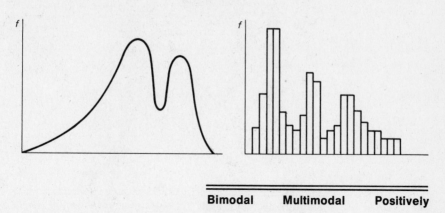

Bimodal   Multimodal   Positively

50. The score that occurs with the most frequency is the mode; hence the mode is totally uninfluenced by extreme scores. The mean is greatly influenced by extreme scores. On the basis of these two statements and the preceding exercises on the median, it is evident that line A indicates the mode since it is not influenced by the extreme scores. Line B is not affected as much as line C, thus it must be the _____. Line C is the _____; it was influenced the most by the extreme scores.

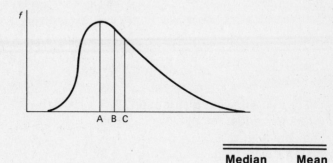

Median   Mean

51. The frequency distribution below is _____ skewed. Line A indicates the _____. Line B indicates the _____. Line C indicates the _____.

The mean of a negatively skewed distribution is located left of the mode. The mean of a positively skewed distribution is located to the _____ of the mode.

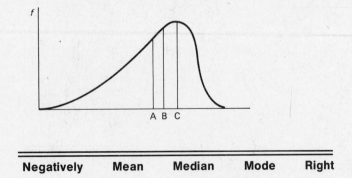

A B C

Negatively	**Mean**	**Median**	**Mode**	**Right**

52. What are the mean, median, and mode of the distribution?

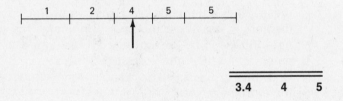

3.4	**4**	**5**

OPEN BOOK QUIZ

1. For the following distribution: 8, 5, 6, 9, 3, 5
   Compute the:
   a. mean _____
   b. median _____
   c. mode _____

2. How could you describe the shape of the following distribution: 2, 2, 3, 3, 3, 3, 4, 4, 5, 6, 8, 9? _____ _____

3. If a small company had mostly low paid employees but one very highly paid employee, what would be the most appropriate mea-

18

sure of central tendency to indicate the salary level of that company?

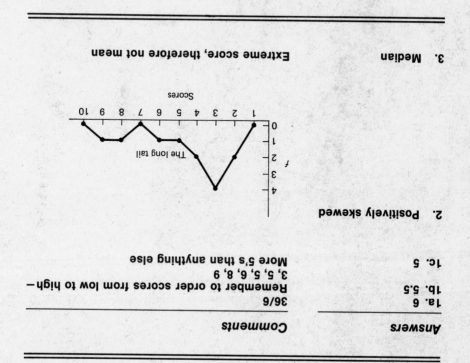

The Normal Curve

53. Let us suppose that each of 260 students lined up in front of signs according to weights. The signs run from left-to-right in order of increasing weight from 135 to 165 pounds. The number of persons in any one line is the frequency of that weight. The

number of 150 pounders is the _____ of 150 pounders.

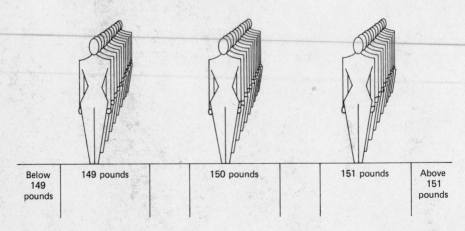

| Below 149 pounds | 149 pounds | 150 pounds | 151 pounds | Above 151 pounds |

Frequency

54. From an airplane, the place where this odd event was occurring might look like the diagram below. Each dot represents a _____.

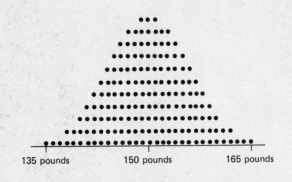

135 pounds   150 pounds   165 pounds

Person (or Student)

55. Assuming that the students are separated from one another by the same amount of space, the *number of cases would be indicated by the area.* For example, with 260 cases, the 26 heaviest

students would occupy the extreme right 10 percent of the crowd. The 13 lightest people would occupy the extreme left _____ percent of the crowd.

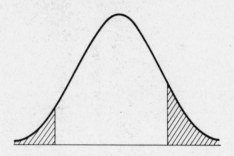

5 Percent

56. If each column of students is represented by a rectangular box, we have our old friend the _____.

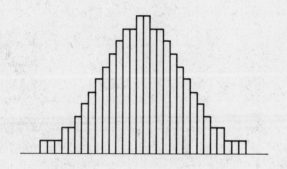

Histogram

57. If we have a very large number of people and use very small weight categories, the irregular steplike curve would become smooth and continuous. The resulting figure approaches a special type of curve called the normal curve. In frequency distributions normality is not associated with small groups of

people but rather with very _____ groups of people.

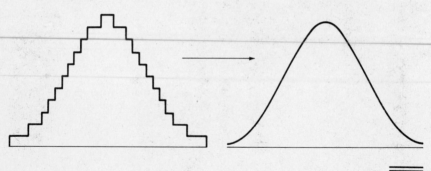

58. In a normal curve (which by definition describes an infinite number of cases) the tails of the curve never touch the baseline. Which curve below could be a true normal curve? _____.

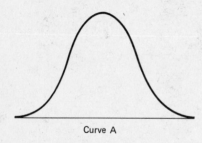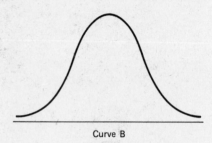

Curve A   Curve B

B

59. It has been found that quantitative data gathered about a variety of natural phenomena, including many mental and social traits, form distributions that, though not precisely normal in distribution, may be closely described by the normal _____.

Curve

60. The distributions of such diverse properties as achievement test scores, IQ, and height and weight of people form approximately _____ _____.

Normal Curves

22

61. The tails of a normal curve recede indefinitely and never touch the abscissa or baseline because the number of cases needed to form a normal curve is _____.

62. When a line approaches infinitely closely to another line but does not touch that line, the lines are said to be asymptotic. The tails of a normal curve are _____ to the baseline.

63. The bell-shaped curve illustrated below approximates what the statisticians call a normal curve. Note the following properties:
   a. It is symmetrical.
   b. The mean, median, and mode have the same value (in this instance, 70).
   c. There are thus an equal number of scores on either side of the mean (central axis).
   d. It is composed of infinitely large numbers of _____.
   e. The tails of the curve are _____ to the abscissa (baseline).

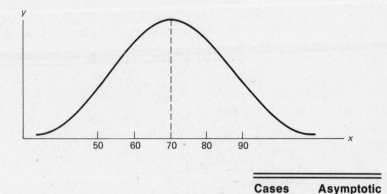

64. Another identifying characteristic of the normal curve is its mathematical construction. There are two points on the normal curve where the curve changes direction from convex to con-

cave. These points are points of inflection (see Graph A). Are the inflection points on Graph B at lines W, lines X, or lines Y?

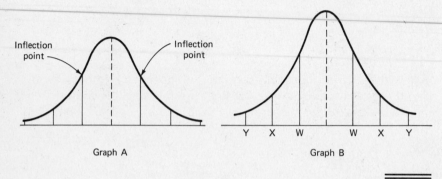

Graph A

Graph B

Lines W

65. Perpendicular lines drawn from the abscissa to the points of inflection may be regarded as marking off *one unit* of distance or deviation from the mean (or central axis). If one uses this distance as a *standard,* a uniform method of dividing the baseline into equal segments (standard deviations) can be established. If the central axis is designated as zero, the line one standard deviation to the right would be plus one and the line one standard deviation to the left would be _____.

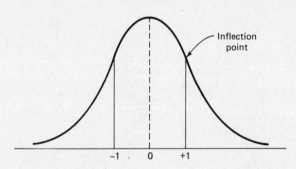

Inflection point

-1 0 $+1$

-1

66. Mathematically, the points -1 and $+1$ are situated one unit of distance or standard deviation from the central axis or values (the mean, median, and mode). These two points are designated

as ±1 (read as plus and minus one). Two units of distance or deviation from the central axis are labeled as +2 and _____.

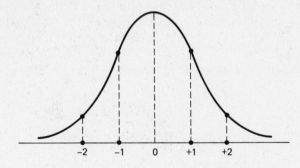

67. Using the unit of distance established by constructing a perpendicular line from the point of inflection to the abscissa as a standard, we can divide the baseline into several equal segments. Since the normal curve is asymptotic with respect to the abscissa, one could divide the baseline into equal parts indefinitely. All segments would be a uniform or standard distance. The unit of distance was established by constructing a perpendicular line from the point of _____ to the abscissa.

68. The proportion of cases beyond ±3 standard deviations from the center of the normal curve is so small that they are generally ignored. It is thus common practice to illustrate only those cases contained between the arbitrary limits of +3 and _____ standard deviations.

69. The number of standard deviations a case is above or below the mean is called the z score for that case. If the case is above the mean, the z score is plus, if below, minus. If a mean was 100 and the standard deviation was 16, a raw score of 116 would be _____ standard deviation(s) _____ the mean and would have a z score of _____.

25

70. The total area under the normal curve may be set to equal 1 or unity. Between the mean and +1 standard deviation to the right of the mean is .3413 of the total area. Thus the area from the mean to +1 standard deviation contains 34.13 percent of the total cases. Since −1 unit of deviation is equal in area to +1 unit of deviation, _____ percent of the total cases lie in the area between a z score of −1 and the mean.

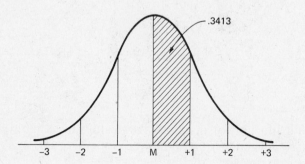

34.13

71. The symmetry and massing of scores around the central values groups little more than $\frac{2}{3}$ (2 × 34.13 percent = 68.26 percent) of the total frequencies between z scores of +1 and −1. If a normal distribution has a total frequency of 1000 scores, approximately 341 scores (34.13 percent of 1000) are located between the mean and a z score of −1 and approximately 341 scores are located between the mean and +1. How many scores are located between z scores of −1 and +1? _____

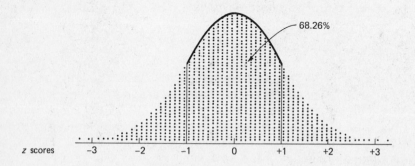

683 (682.6)

26

72. In this frequency distribution the z scores, −1 and +1, mark off the middle _____ percent of the total scores. They occur at the scores of 40 and _____.

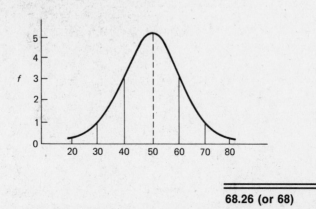

73. Although the normal curve extends indefinitely to the left and to the right, the end points of the curve approach the baseline so closely that over 95.44 percent (see graph below) of the area or frequencies are included between the limits −2 and +2 and 99.74 percent of the cases are included between the limits −_____ and +_____.

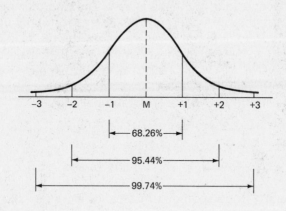

74. The percentage of cases contained between the mean (central axis) of a normal curve and a z score of +3 is 49.87 percent

(one-half of 99.74 percent). The percentage of cases between the mean of a normal curve and a z score of −2 is _____ percent.

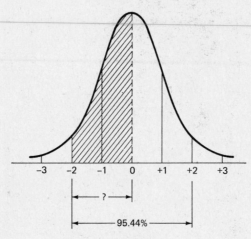

75. As we stated before, for practical purposes the limits of the frequencies of the normal curve rarely exceed those of ±3 standard deviations from the mean. The approximate twenty-six-hundredths of 1 percent (.0026) of the total cases occurring outside the limits of ±3 is so slight that unity is generally assumed to be within these limits. Approximately thirteen hundredths of 1 percent (.0013) of the total cases extend beyond +3 and approximately _____ hundredths of 1 percent of the total cases extend beyond −3.

76. Though the percentage of cases is very small and insignificant at a considerable distance from the mean (beyond ±3), the proportion of frequencies approaches zero but never equals zero. The reason is that the normal curve is _____ with respect to the abscissa or baseline.

77. Since the proportion of the total cases that exist beyond the limits of ±3 is so slight, it may be plausible to treat data as if 100 percent of the total cases fall within ±3 standard deviations. If one makes this assumption, the percentage of cases is rounded off to the nearest whole percent. Thus (note graph below) the percentages of cases from the mean to +1, +2, and +3 are 34 percent, 48 percent, and 50 percent, respectively. The percentages of cases *from the mean* to −1, −2, and −3 standard deviations are _____ percent, _____ percent, and _____ percent, respectively.

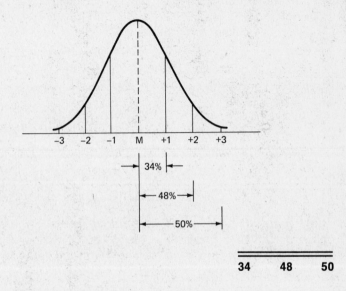

34   48   50

78. The percentage of cases below the mean is _____ percent.

50

79. The percentage of cases between the mean and +1 is _____ percent of the total cases.

34

80. The percentage of cases below a *z* score of +1 is 84 percent (50 percent plus 34 percent) of the total cases and the per-

centage of cases above +1 is _____ percent of
the total cases. The percentage of cases below −1 is _____
percent.

81. In relation to the scores on the graph below, about what percentage of cases lie below 43? _____ Between
43 and 57? _____ Below 57? _____
Above 50? _____ Below 36? _____
Total of below 36 and above 50? _____

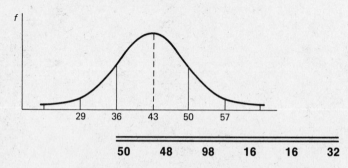

People do not usually memorize the proportions above or
below particular *z* scores but make use of tables such as Table 1
(an asterisk preceding a value means that the value is exact).

82. What proportion of the cases in Table 1 would have a *z* score
equaling or exceeding 1.645? _____

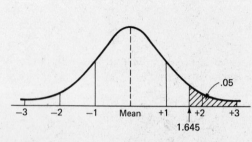

(The .05 was obtained by going down the column marked *z
Score* to the desired *z* (1.645), then, since we want the proportion
above the *z*, we go to the first column finding .05 or 5%.)

30

Table 1   (For a more complete table, see pages 140–146.)

Normal Distribution

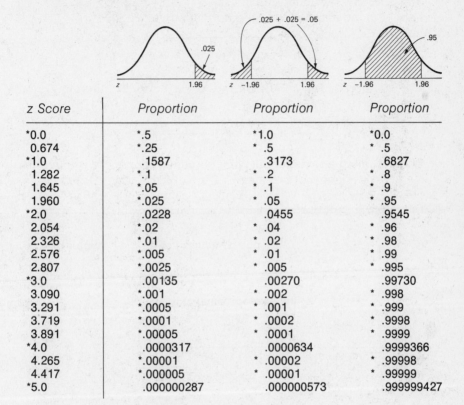

z Score	Proportion	Proportion	Proportion
*0.0	*.5	*1.0	*0.0
0.674	*.25	* .5	* .5
*1.0	.1587	.3173	.6827
1.282	*.1	* .2	* .8
1.645	*.05	* .1	* .9
1.960	*.025	* .05	* .95
*2.0	.0228	.0455	.9545
2.054	*.02	* .04	* .96
2.326	*.01	* .02	* .98
2.576	*.005	* .01	* .99
2.807	*.0025	* .005	* .995
*3.0	.00135	.00270	.99730
3.090	*.001	* .002	* .998
3.291	*.0005	* .001	* .999
3.719	*.0001	* .0002	* .9998
3.891	*.00005	* .0001	* .9999
*4.0	.0000317	.0000634	.9999366
4.265	*.00001	* .00002	* .99998
4.417	*.000005	* .00001	* .99999
*5.0	.000000287	.000000573	.999999427

OPEN BOOK QUIZ

1. The distribution of dots below approximates a _____
   distribution.

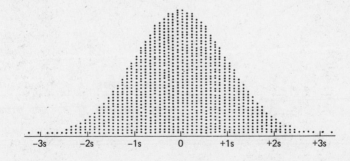

31

2. The scores below a z of -1.96 and above a z of $+1.96$ would make up what proportion of the total scores? _____

3. What proportion of scores would be between z scores of -1.645 and $+1.645$? _____

4. Label the indicated place on a normal curve.

_____ _____

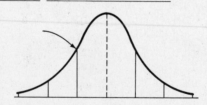

5. Without using a table, about what percent of cases are within one standard deviation of the mean? _____ percent.

Answers	Comments
1. Normal	See Frame 63.

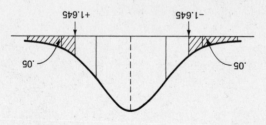

2. .05 or 5 percent

.025 + .025 = 0.5

3. .90 or 90 percent

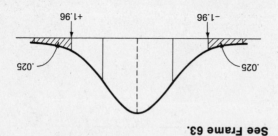

If .10 outside, then .90 inside.

4. Inflection point — See Frame 64.

5. 68 percent — See Frames 70 and 71.

Variability

83. Descriptions of groups by frequency distributions, central tendency, and normality have been discussed. Another way of describing a group is to have some index of how much variability exists. Consider the height of the two groups of people below. Both groups have a mean and median of 6 feet but the more variable is Group _____.

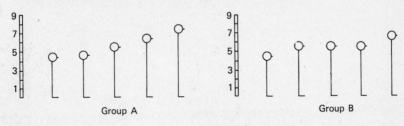

Group A   Group B

A

84. One common measure of variability is the range. The range of a set of scores is the distance between the midpoints of the lowest and highest scores. To find the range, subtract the lowest score from the highest score. The range of Group A with heights of $4\frac{1}{2}$, 5, 6, 7, $7\frac{1}{2}$ is $7\frac{1}{2}$ minus $4\frac{1}{2}$ or 3. The range for the less variable Group B with heights of 5, 6, 6, 6, 7 is _____.

2 (or 7 Minus 5)

85. If the normal curves below, in which the vertical deviation lines are one standard deviation apart, represent large populations, which curve represents the most variable group? _____

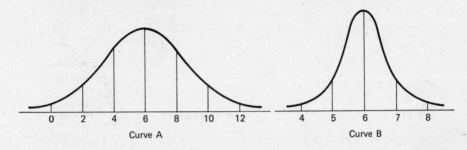

Curve A   Curve B

Curve A

33

86. The size of the standard deviation is the other major index of variability. In the diagram below, 29 differs from 43 by two _____ _____.

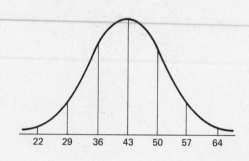

22   29   36   43   50   57   64

Standard Deviations

87. When members of a group deviate very little from each other, the standard deviation is very small. The reverse is true for highly variable groups. Consequently, the variability or diversity of two groups can be compared by the relative size of their

_____ _____.

Standard Deviations

88. The capital letters SD are used when referring to the standard deviation of a sample of a population. It is common practice to symbolize the standard deviation by the small Greek letter sigma (σ) when referring to the population values. To abbreviate a sample's standard deviation, one could use the capital letters _____. To abbreviate the standard deviation for a population, use the small Greek letter _____.

SD σ (or Sigma)

89. Suppose that a population of scores is distributed so that the mean is 40 and the distance of 12 points is 1σ (read as one

standard deviation). The students one standard deviation above the mean received the score of _____.

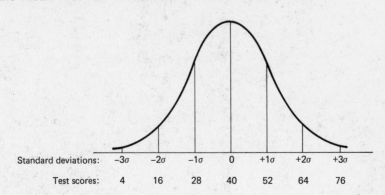

Standard deviations:	−3σ	−2σ	−1σ	0	+1σ	+2σ	+3σ
Test scores:	4	16	28	40	52	64	76

$$\underline{\underline{52}}$$

90. What is the difference, in score points, between the scores of +1σ and −2σ of the above illustration? _____

$$\underline{\underline{36 \ (3 \times 12)}}$$

91. The standard deviation is a kind of average of all the deviations from the mean score. The amount a score (X) deviates from the mean ($\overline{X}$) is symbolized by the small letter x; that is, $X - \overline{X} = x$. Give the symbols for the following:
Raw score _____
Mean _____
Deviation _____

$$\underline{\underline{X \qquad \overline{X} \qquad x}}$$

92. To calculate a standard deviation, the deviations from the mean have to be squared. To square a number, you multiply it by itself. For example, 5 squared, or 5^2, $= 5 \times 5 = 25$; $2^2 =$ ____.

$$\underline{\underline{4}}$$

35

93. A negative value times a negative value gives a positive value; therefore, $(-4)^2$ or $-4 \times -4 = 16$. Complete the following:

$$(-7)^2 = \underline{\hspace{3cm}}$$
$$(-3)^2 + (-1)^2 + (-1)^2 + 5^2 = \underline{\hspace{3cm}}$$

$\overline{\overline{49 \qquad 36}}$

94. The opposite of squaring a number is taking a square root. The square root of 25 or $\sqrt{25} = 5$; $\sqrt{16} = \underline{\hspace{3cm}}$.

$=$
4

95. When some arithmetic occurs inside a square-root sign, work the arithmetic before taking the square root.

$$\sqrt{1+3} = \sqrt{4} = 2$$
$$\sqrt{\frac{5^2+7}{2}} = \sqrt{\frac{32}{2}} = \sqrt{16} = \underline{\hspace{2.5cm}}$$
$$\sqrt{\frac{(-3)^2 + (-1)^2 + (-1)^2 + 5^2}{4}} = \sqrt{\frac{?}{4}} = \sqrt{?} = \underline{\hspace{2cm}}$$

Note: When we take the square root of a number, both positive and negative roots occur; but we are concerned only with the positive roots.

$\overline{\overline{4 \qquad \sqrt{36/4} = \sqrt{9} = 3}}$

96. Squaring, dividing, and taking the square root are used in solving the formula for the standard deviation: $SD = \sqrt{\Sigma x^2/N}$. The symbol $\sqrt{}$ directs a person to take the square _____.

$\overline{\overline{\text{Root}}}$

97. The formula for the standard deviation, $\sqrt{\Sigma x^2/N}$, can be executed in six steps:
   a. Compute the mean $(\overline{X})$.
   b. Subtract the mean from each score to find the deviations from the mean $(X - \overline{X} = x)$.
   c. Square each deviation from the mean (x^2).

d. Add the squares of the deviations (Σx^2). This is
   called the sum of squares and often symbolized as SS.
e. Divide the SS by the total number of scores (SS/N).
f. Take the square root of SS/N, ($\sqrt{SS/N}$).
The steps above illustrate why the standard deviation is defined
as the square root of the average of the squared deviations from
the _____.

══
Mean

98. To see how to calculate a standard deviation, let's assume a dis-
    tribution of 0, 2, 2, 8. The mean ($\overline{X}$) is 3.

$$X - \overline{X} \text{ or } x \text{ (deviations)} = (0\text{–}3),\ (2\text{–}3),\ (2\text{–}3),\ (8\text{–}3)$$
$$= -3,\ \ \ \ -1,\ \ \ \ -1,\ \ \ \underline{\quad}$$
$$x^2 \text{ (squared deviations)} = 9,\ \ \ \ \ 1,\ \ \ \ \ 1,\ \ \ \ \underline{\quad}$$
$$\text{SS} = \Sigma x^2 \text{ (sum of squared}$$
$$\text{deviations)} = 9\ +\ 1\ +\ 1\ +\underline{\quad},\ \ \underline{\quad}.$$

══════
5   25   25, 36

99. We have found the sum of the squared deviations (SS) to be 36.
    With an N of 4, the average of the squared deviations from the
    mean is SS/$N = 36/4 = 9$. The square root of the average of the
    squared deviations from the mean equals 1 standard deviation.
    That is, SD $= \sqrt{SS/N} = \sqrt{36/4} = $ _____.

══
3

100. Calculate the standard deviation for the distribution 1, 3, 4, 5, 7.
    a. $\overline{X} = 4$
    b. $(X - \overline{X})$ or $x = (1\text{–}4),\ (3\text{–}\underline{\ \ }),\ (\underline{\ \ }\text{–}\underline{\ \ }),\ (\underline{\ \ }\text{–}\underline{\ \ }),\ (\underline{\ \ }\text{–}\underline{\ \ })$
    c. $x = $ $\underline{\ \ },\ \ \underline{\ \ },\ \ \underline{\ \ },\ \ \underline{\ \ },\ \ \underline{\ \ }$
    d. $\Sigma x^2 = $ SS $= $ $\underline{\ \ } + \underline{\ \ } + \underline{\ \ } + \underline{\ \ } + \underline{\ \ }$
    e. SS/$N = $ ___
    f. SD $= \sqrt{SS/N} = $ ___.

══════
4, (4–4), (5–4), (7–4)   −3, −1, 0, 1, 3   9, 1, 0, 1, 9   4   2

Sum of Squares: Practical Computation

Not only is the squaring of frequently decimal values in computing the SS tedious when N is large, but the process involves two passes with a calculator or computer; the first to determine the mean, the second to compute and square differences and sum those squares. It is all a bit awkward. There is, fortunately, an algebraically equivalent formula that, while it appears more foreboding at first, is much faster and more convenient in actual practice.

This formula, to be explained shortly, for SS is:

$$SS = \Sigma X^2 - \frac{(\Sigma X)^2}{N}$$

Notes	Example Problem			Practice Problem		
		X	X^2		X	X^2
101. Compute ΣX as before, but in addition square each score (X^2) and sum up the squares (ΣX^2).	Don	0	0	Ima	0	___
	Ray	4	16	Lil	2	___
	Jan	2	4	Dot	2	___
	May	4	16	Hal	3	___
	Joy	6	36	Sue	8	___
	Jim	4	16			
	Sam	4	16			
	Fay	6	36			
	Art	6	36			
		$\Sigma X = 36$			$\Sigma X = $ ___	
			$\Sigma X^2 = 176$			$\Sigma X^2 = $ ___

	0
	4
	4
	9
	64
15	81

102. Square the ΣX. Note that $(\Sigma X)^2$ is the symbol. Do not confuse this	$(\Sigma X)^2 = 36^2$ $= 1296$	$(\Sigma X)^2 = ($_____$)^2$ $= $_____

with ΣX^2. For ΣX^2
we square first
and then sum;
for $(\Sigma X)^2$ we sum
first and then
square.

$$\frac{15}{225}$$

	Notes	*Example Problem*	*Practice Problem*
103.	Divide $(\Sigma X)^2$ by N. The result is called the correction factor (CF).	$CF = (\Sigma X)^2/N$ $= 36^2/9$ $= 1296/9$ $= 144$	$CF = \underline{\hspace{2cm}}/\underline{\hspace{2cm}}$ $= \underline{\hspace{2cm}}$

$$\frac{225/5}{45}$$

	Notes	Example Problem	Practice Problem
104.	Subtract CF from ΣX^2. The result is the corrected sum of squares (SS).	$SS = \Sigma X^2 - CF$ $= 176 - 144$ $= 32$	$SS = \underline{\hspace{2cm}} - \underline{\hspace{2cm}}$ $= \underline{\hspace{2cm}}$

$$\frac{81 - 45}{36}$$

The "correction" the CF makes is to compensate for using raw scores rather than deviation scores (i.e., $X - M$). When deviation scores are used, since their mean and sum would equal 0.0, the CF drops out.

105. Subtracting a constant from or adding a constant to all the raw scores of a distribution does not change the value of the standard deviation. If the standard deviation of 50, 52, 52, 58 is 3, the standard deviation of 50–50, 52–50, 52–50, 58–50 or 0, 2, 2, 8 is

_____.

$$\overline{3}$$

106. The range is easier to understand and easier to calculate than the standard deviation but it has some serious disadvantages. Not much else can be done with the range. The standard deviation (and its square called the variance) is used in many kinds of statistical analysis. The measure of variation having the greater versatility is the _____.

Standard Deviation

107. The size of the range depends a good deal upon the size of the sample. There is more chance of simultaneously drawing a very high score and a very low score when the sample is larger. Consequently, range generally increases with an increase in the size of the _____.

Sample

108. Because all the scores are used in computing the standard deviation while only two scores (the highest and lowest) are used in computing the range, the standard deviation is much more stable than the range. The more stable measure of variability is the _____.

Standard Deviation

109. For example, a sample of 20 scores could be drawn at random from a population of 200 scores. The standard deviation and the range could now be calculated and the 20 scores returned to the population pile. If this process were repeated many times, the standard deviations would vary in size much _____ than would the range.

Less

40

1. Compute the SS for the distribution: 2, 2, 6, 6. _____
2. Compute the SD for the distribution 2, 2, 6, 6. _____
3. Based on question 2, and without doing any further calculations, what would be the SD for the distribution 3, 3, 7, 7? _____
4. Compute the range for the distribution 2, 2, 6, 6. _____
5. Which distribution below would have the larger SD? _____

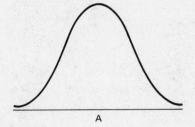

A

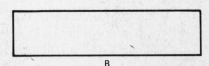

B

Answers	Comments
1. 16	$\Sigma X^2 = 2^2 + 2^2 + 6^2 + 6^2 = 80$
	$(\Sigma X)^2/N = 16^2/4 = 64$
	$SS = 80 - 64 = 16$
2. 2	$SD = \sqrt{SS/N} = \sqrt{16/4} = 2$
3. 2	Adding a constant (1 in this case) to each score does not change the SD.
4. 4	$6 - 2$
5. B	Scores of A are mostly bunched together.

Relationship: Correlation Coefficient

110. Many variables or events in nature are related to each other. As the sun rises, the day warms up; as children age, they think more complexly; and persons bright in one area tend to be bright in others. Such relationships are called correlations. The relationship of one variable to another is known as a _____.

Correlation

111. If the river rises when it rains, the two events are said to have a positive correlation. That is, when an increase in one variable coincides with an increase in another variable, the two variables have a _____ correlation.

Positive

112. Altitude and air pressure have a negative correlation. The greater the altitude, the less the air pressure. When an increase in one variable coincides with a decrease in another variable, the two variables have a negative correlation. With children, bedwetting and age usually have a _____ _____.

Negative Correlation

113. When there is a high correlation between two variables, we can predict the values for one variable from those of the other. If there is a high positive correlation between drownings and ice cream sales we can predict that as ice cream sales increase the number of drownings will _____.

Increase

114. We can predict the occurrence of one event from another event, but we cannot say that one event *causes* the other event. There is a positive correlation between the number of drownings per

day and ice cream sales, but drownings do not cause the ice cream sales or vice versa. A third variable, heat, is probably the _____ of both events.

115. There must be a common link between the sets of variables being correlated. If two tests are to be correlated, the same person or persons who take both tests must be matched on related variables. Could one correlate the performance of a fifth-grade class on test A with the performance of another *unmatched* fifth-grade class on test A? _____.

116. The most common numerical measure of correlation is the product moment correlation coefficient. This is commonly symbolized by the small letter *r*. Whenever we see *r*, it symbolizes the

_____ _____ _____

_____.

117. The value *r* measures the degree to which the relationship between two variables can be represented by a *straight* line. Which of the variables in the diagrams below, *X, Y,* or *Z,* has the strongest *r* with the variable *A*? _____.

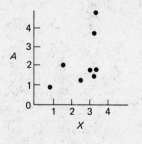 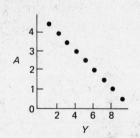 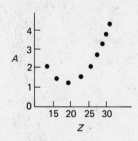

43

118. The value *r* ranges from +1.00 for a perfect positive linear (straight line) relationship, through 0.00 for no linear relationship, to −1.00 for a perfect _____ linear relationship.

119. We have learned that when high values on one test tend to go with low values on another test the tests are negatively correlated. The algebraic sign of minus (−) indicates a negative correlation or an inverse relationship. The algebraic sign of plus (+) indicates a positive correlation or a direct relationship. Which diagram illustrates inversely related variables having an *r* = −1.00? _____

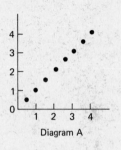 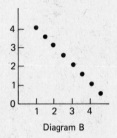

Diagram A   Diagram B

120. A correlation of −1.00 is as strong a correlation as +1.00. The algebraic sign (+ or −) of the correlation coefficient indicates the direction of the relationship (whether direct or inverse). It is the absolute size of *r* that indicates the degree of strength or closeness of the relationship. Is an *r* of −0.80 higher or lower than an *r* of +0.65? _____

121. In the real world it is more common than not for the value of *r* to be much lower than −1.00 or +1.00. The closer the dots ap-

proach a straight thin line, the higher the *r*. Which of the variables (*X*, *Y*, or *Z*) has the highest *r* with the variable *A*? _____ Which the lowest? _____

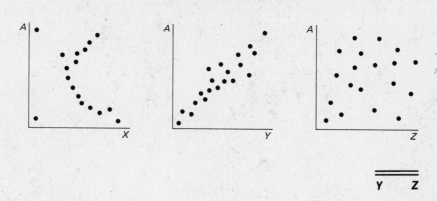

122. To get a feel for the closeness of relationship indicated by various sizes of *r* (all positive for easy comparison), compare the following graphs, and estimate the *r* of Graph E. _____

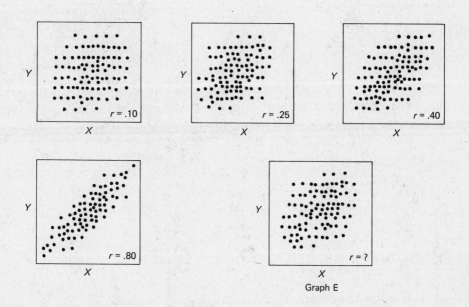

Graph E

In computing the correlation coefficient, a value similar to the SS must be found; it is the sum of products (SP).
The formula for the SP, to be explained below, is:

$$SP = \Sigma XY - \frac{(\Sigma X)\,(\Sigma Y)}{N}$$

Notes	Example Problem			Practice Problem		
	X	Y	XY	X	Y	XY

123. X represents the scores on the visual perception test. Y is the score on the reading test. Each Y score is multiplied by its corresponding X score. The products are then summed (ΣXY).

Example Problem:

$0 \times 2 = 0$
$4 \times 2 = 8$
$2 \times 3 = 6$
$4 \times 4 = 16$
$6 \times 4 = 24$
$4 \times 7 = 28$
$4 \times 8 = 32$
$6 \times 7 = 42$
$6 \times 8 = 48$
$\Sigma X = 36$
$\Sigma Y = 45$
$\Sigma XY = 204$

Practice Problem:

X	Y	XY
0	3	__
2	1	__
2	2	__
3	3	__
8	1	__

$\Sigma X = $ __ $\Sigma Y = $ __ $\Sigma XY = $ __

		0
		2
		4
		9
		8
15	10	23

124. Note the similarity of the formula for SP to that for SS. With SS it was the sum of all the X times X (i.e., itself) and then (ΣX) times (ΣX). With SP it is the sum of all the X times their respective Y and then (ΣX) times (ΣY).

Example Problem:

$SP = \Sigma XY - (\Sigma X)\,(\Sigma Y)/N$
$= 204 - (36)\,(45)/9$
$= 204 - 180 = 24$

Practice Problem:

$SP = $ __ $-$ __ $/$ __

$= $ __

$23 - (15)\,(10)/5$
-7

Notes	Example Problem		Practice Problem	
	Y	Y^2	Y	Y^2

125. The SS for Y (SS_y) will be needed and is computed in the same way as SS_x.

Notes	Example Problem		Practice Problem	
	Y	Y^2	Y	Y^2
	2	4	3	____
	2	4		
	3	9	1	____
	4	16		
	4	16	2	____
	7	49		
	8	64	3	____
	7	49		
	8	64	1	____
	45	275		____

$$SS_y = \Sigma Y^2 - (\Sigma Y)^2/N$$
$$= 275 - 45^2/9$$
$$= 50$$

$SS_y = \underline{} - \underline{}/\underline{}$

$= \underline{}$

$$\begin{array}{c} 9 \\ 1 \\ 4 \\ 9 \\ \underline{1} \\ 24 \end{array}$$

$24 - 10^2/5$
4

126. The formula for r is:

$$r = \frac{SP}{\sqrt{SS_x SS_y}}$$

Notes	Example Problem	Practice Problem
Collecting the needed statistics from previous calculations.	$SP = 24$ $SS_x = 32$ $SS_y = 50$	$SP = \underline{}$ (see Frame 124) $SS_x = \underline{}$ (see Frame 104) $SS_y = \underline{}$ (see Frame 125)

-7
36
4

	Example Problem	Practice Problem

127.

$$r = SP/\sqrt{SS_x SS_y}$$

$$= 24/\sqrt{(32)(50)}$$ $$= \underline{\quad}/\sqrt{(\quad)(\quad)}$$

$$= 24/\sqrt{1600}$$ $$= \underline{\quad}/\sqrt{\underline{\quad}}$$

$$= 24/40$$ $$= \underline{\quad}/\underline{\quad}$$

$$= .60$$ $$= \underline{\quad}$$

$$-7/\sqrt{(36)(4)}$$

$$-7/\sqrt{144}$$

$$-7/12$$

$$-.58$$

128. A correlation coefficient is not a direct measure of the percentage of relationship between two variables. One cannot say that a correlation of +0.90 is three times as close a relationship as +0.30, but merely that it indicates a much higher degree of relationship. The correlation values or coefficients of correlation are not measurements on a scale of equal units. Are two tests having an r of +0.40 twice as related as two tests having an r of +0.20? _____

No

129. If the correlation values are not measurements on a scale of equal units, then an *increase* in correlation, for example, from +0.30 to +0.50, is *not* equal to the same amount of *increase* on another correlation, for example, from +0.60 to +0.80. Each of the two correlations experienced an _____ in the closeness or similarity of the variables they were measuring but not necessarily to the same extent or degree.

Increase

130. Whether a correlation is considered high or not depends on what we are correlating. Some predictions do not have to be very precise to be of important use, so as a result even a small amount of correlation is noteworthy. Ignoring the *use* of a

correlation, an overall "rule of thumb" for judging correlation size is to consider an r of 0.70 to 1.00 (either + or −) as a high correlation and an r of 0.20 to 0.40 as a relatively low correlation. Disregarding use, how would you describe a correlation of .35? _____

Low (or Relatively Low)

OPEN BOOK QUIZ

1. If grades tend to raise as class attendance increases, one could say that grades and attendance are _____ _____.

2. Which of the following represents the highest correlation?

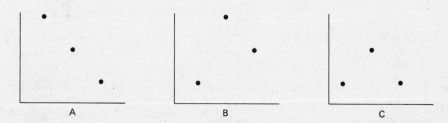

A   B   C

3. Comment on the assertion that a correlation rise in relationship as r increases from .5 to .6 is equal to that when r increases from .7 to .8. _____ _____

4. Compute r for the following pairs of scores. $r =$ _____.

	X	Y
Ima	1	1
Lil	2	3
Dot	3	2

5. Compute r for the following pairs of scores. $r =$ _____.

	X	Y
Don	1	2
Ray	2	3
Joy	3	1

Answers	Comments
1. Positively correlated	
2. A	See Frame 117
3. Not true	See Frame 129
4. +.5	$\Sigma XY = 13 \quad \Sigma X = \Sigma Y = 6 \quad N = 3$ $SP = 13 - (6 \times 6/3) = 1$ $\Sigma X^2 = \Sigma Y^2 = 14 \quad SS = 14 - (6^2/3) = 2$ $r = 1/\sqrt{2 \times 2} = 1/2 = .5$
5. -.5	$\Sigma XY = 11$ All else is same as Question 4. $SP = 11 - (6^2/3) = -1$ $r = -1/\sqrt{2 \times 2} = -1/2 = -.5$

Inferential Statistics

131. The statistical ideas presented so far are useful for *describing* central tendencies, degree of variability, relative standing, and correlation with available data. These were, for the most part, *descriptive* uses of statistics. That branch of statistics is called _____ statistics.

Descriptive

132. It is frequently useful to be able to predict the future or learn things about unseen persons. The branch of statistics that allows us to do this goes beyond description and *infers* unknown data from known data. Because it is inferential rather than descriptive, it is called _____ _____.

Inferential Statistics

133. A *population*, as used in inferential statistics, is *all* of something about which an inference is to be made. It might be all the senior high school students in the United States, all the wheat in a ship's hold, all high school seniors having IQs of 128, English grades of A and who are female, etc. A *sample* is a portion of some *population*, such as some of the seniors at Centerville High, a bucket of wheat from a ship's hold, etc. It should be much easier to gather data about a _____ than about a _____.

Sample   Population

134. What is called a population depends upon the researcher's objectives. If he wishes to infer something about the students of Centerville High, the school becomes the population and Mr. Smith's class might be the sample. If we wished to infer something about a bucket of wheat, 100 grains might be the sample and the bucket the _____.

Population

135. If the freshman class at a university were measured on the basis of some variable, the freshmen might be considered by a researcher as a sample of (a) all freshmen ever to attend that university; (b) present freshmen at all universities; (c) men and women around the age of 18; etc. In general the students may be considered as a _____ of various _____.

Sample   Populations

136. *Sample* and *population* are often used by statisticians to refer, not to people or things, but to data. In this sense, the sample would not be the freshman class at a university but perhaps the SAT scores of these freshmen. The population might be the SAT _____ of all freshmen ever to attend that university.

Scores

137. In the same sense, a single test score by a student could be viewed as a sample of all his scores if he were to take the same test an infinite number of times (each time knowing only what he knew the first time he took the test). This hypothetical infinite set of scores would comprise the _____.

Population

138. The easiest type of sample to understand is the *simple random sample*. (There are other types of valid random samples besides the simple variety, but a complete listing would go beyond the scope of this brief discussion.) For a sample to be a simple random sample, it must be true that any single observation or score in the population (or observations or scores) has the same chance as any other single observation or score (of that population) of being included in the sample and the observations must be independent of each other. Would the heights of all the freshmen at a university in one year be a simple random sample of the heights of all freshmen ever to attend that university? _____

Probably Not

139. No matter how sophisticated the statistical techniques used, one would be inviting gross error if one presumed test results to be valid for a population of *all* who attempt suicide when one's sample consisted only of those who were unsuccessful. Such a sample is called a biased sample. To be representative of an entire population a sample must *not be* _____ in regard to that population.

Biased

140. If one tried to predict the results of a presidential election by picking names out of country club member lists and questioning the people selected, one would be working with a(n) _____ sample of the United States population.

Biased

141. A value for a sample, such as a mean or SD, is called a *statistic*. The corresponding value for a population, such as a mean or σ, is called a *parameter*. In inferential statistics, one uses statistics to estimate _____.

Point Estimation

142. There are three general kinds of inferences commonly drawn about parameters: *point estimates, confidence intervals*, and *significance tests*. In point estimation we seek the best single value that can be used to estimate a parameter. The best estimate of the mean or median of a population is the mean or median of the sample. If a sample had a mean of 120, the best estimate of the population mean would be _____.
An estimation of a population mean or median based on the sample's mean or median is a point _____.

143. For reasons beyond this brief inquiry into statistics, a population's standard deviation is estimated by using $N - 1$ instead of N as a divisor. If the sum of squares of *deviations* from the mean for a sample with an N of 5 were 36, the population standard deviation would be estimated at $\sqrt{36/(5-1)}$, or _____.

144. In Frames 101 to 104, an example problem was worked that had an N of 9; the SS turned out to be 32. Using these 9 as a sample, estimate the standard deviation of the population.

_____

145. A practice problem was in the same frames. Go back and briefly review how you got the SS for it. Observe the *N*. Now estimate the standard deviation of the population of which these 5 people were a sample. _____

$$S = \text{Estimate of SD} = \sqrt{\frac{36}{5-1}} = \sqrt{9} = 3$$

Confidence Intervals

146. An interval within which a parameter would most probably fall is called a *confidence interval.* (Frames 231 through 240 develop this concept.) If one concluded from a sample that there was a 95 percent (i.e., 95 out of 100 or 19 out of 20) chance that some population had a mean IQ somewhere between 111 and 130, the interval from 111 to 130 would represent the 95 percent _____ interval.

Confidence

Significance Testing

147. If one were to draw several samples at random from a single population, these samples might very well have different values (e.g., 5 different 100 grain samples of wheat from a ship's hold might have different mean kernel lengths). Assuming that the samples are unbiased, these sample fluctuations can be attributed to chance. If one were to examine thousands of such samples (and determine the mean kernel length of each sample as well as of the entire shipload), one would find that the means of these samples formed a near normal curve. Most of the means would be clustered very close to the population mean. Fewer and fewer means would occur at each gradation of deviation from the population mean, just as in any normal distribution. Even the extreme deviations of the sample means from the population mean are presumably due to _____.

Chance

148. Although the sample means varied, we know that the population means for all the samples of wheat were identical as they were drawn from the same population. In the behavioral sciences, we usually do not know the population values but can only _____ the most likely population values from the sample values.

Infer

149. Suppose that one were comparing two textbooks for teaching concepts of statistics, say a programmed text and an orthodox text. A sample of students is drawn from a population and each student is randomly assigned to one of two experimental groups. One group would use the programmed text and the other would use the orthodox text. Afterward, their achievement would be measured on the same test. In all likelihood, whether there was a difference in the two textbooks or not, the test scores for the two groups would have different means. Thus, one would wish to know: "Is the difference in test values attributable to a difference in textbooks or to the chance fluctuation of sample means about some common population _____?"

Mean

150. *Significantly different* as in "These means are significantly different," indicates that the two sample means are probably not drawn from a common population (as far as the characteristic being measured goes). If two different methods of teaching yielded achievement score means of 129 and 130 respectively and standard deviations of 10 and 12 respectively, would you estimate that the two samples differed significantly from each other? _____

No

Note: When we say that a correlation between two variables (as determined from a sample) is *significant*, we mean that it is *not* very probable that the sample is a random sample drawn from a population in which two variables have a correlation of zero.

Null Hypothesis

151. In order to determine whether means are significantly different from each other, statisticians often employ the strategy of testing the hypothesis (called the *null hypothesis*) that these means come from the same population; in other words, they test to see whether the mean differences can be explained as chance fluctuation about a common mean. If a statistical test showed the probability to be quite low that the sample means are fluctuating about a common mean (traditionally below 5 chances in 100 or .05, where .00 means no possibility and 1.00 means absolute or 100 percent certainty), then one might appropriately reject the _____ hypothesis.

Null

152. For example, consider two matched classes, one receiving teaching method A and the other B. A null hypothesis might be as follows: "The amount of learning achieved by students following method A is the same as or equal to that achieved by students following method B." If the data forced rejection of this hypothesis, one could say that the achievement levels obtained under method A and B differed _____.

Significantly

153. Even when the null hypothesis (often symbolized H_0) *is* true, particular pairs of samples would have differences between their means ranging from nothing to quite large values. If a particular experiment yields a pair of means at least as far apart as the 5 percent most divergent pairs of values expected when the H_0 *is* true, the results are said to be significant at the 5 percent or .05 level. If an experimenter chose to call pairs significantly different only if they were as divergent as the most extreme 1 percent of theoretical pairs would be if they were from the same population, he would be working at the _____ _____ _____.

1 percent (or .01) Significance Level

56

154. Since the significance level refers to the maximum chance for rejecting the null hypothesis *when* the null hypothesis is true, an investigator who restricts his investigations to the obvious (such as setting null hypotheses as: men do not differ from women in physical strength, ninth graders do not differ from fifth graders in knowledge of science, there is no relationship between air temperature and ice cream sales, etc.) will almost always find significant differences and, as the null hypothesis is so rarely true in his experiments, will err much less than 5 percent of the time in his rejection of the _____ _____.

Null Hypothesis

155. Does using the 5 percent level mean that one will be wrong on the average of 1 out of every 20 rejections of the null hypothesis? _____

No

156. Since the 5 percent level is more "generous" in accepting statements as true than is the 1 percent level, it is less likely to miss real differences or effects existing in nature. If one didn't particularly mind being wrong when saying a correlation was significant but was very anxious not to miss any relationships that did exist, would it be better to use a 1 percent level or a 5 percent level? _____ _____

5 Percent Level

Significance of a Correlation Coefficient

You might wish to know if a correlation is large enough so that one could feel pretty certain that the relationship was not just a fluke. That is, is the r of a sample large enough that it is very improbable (say less than 5 percent) that the population r is 0.0? To put this yet another way: is the correlation significant at the .05 level? It's easy to find out. Just enter Table 2 with the number of pairs minus 2 ($N - 2$) and, in order to reject the null hypothesis of no relationship in the population, your r has to be as large or larger than the table value (either in a positive or negative direction).

Table 2   (For a more complete table, see page 135.)

.05 Points for the Distribution of r

$N - 2$ (Degrees of freedom)	r
1	.997
2	.950
3	.878
4	.811
6	.707
8	.632
12	.532
20	.423
30	.349
60	.250
120	.179

157. In the earlier calculations of r, a sample of 9 pairs of scores was found to have an r of .60. Would this indicate a probable positive correlation in the population? _____

No. With df $= 9 - 2 = 7$, the r needed would be somewhere between .632 and .707. With our computed r of only .60 we cannot reject luck as the explanation for the correlation in the sample.

OPEN BOOK QUIZ

1. Researchers analyze _____ from samples in order to make inferences about the _____ of _____.
2. If a SS for a sample of 3 cases was 8, estimate the SD of the population.
3. If one were to conclude that there was a 95 percent chance that a population r lay between .6 and .8, he would have established a _____ interval.
4. If one is working at the .05 significance level, it means he had a maximum probability of .05 of rejecting the _____ _____ when it is true.
5. Comment on the significance of an r of .5 obtained from 22 pairs of scores.

Answers	Comments
1. Statistics Parameters Populations	See Frames 133 to 137
2. 2	$\sqrt{8}/(3-1) = \sqrt{8}/2$
3. Confidence	See Frame 146
4. Null hypothesis	See Frames 151 to 152
5. Significant at .05 level	With df $= 22 - 2 = 20$, r needed is .423. Therefore r of .5 is significant (Table 2 preceding Frame 157)

Analysis of Variance

158. It is common to test to see if the separate means of *several* groups differ significantly from each other. In such an instance *all* the cases in *all* the groups averaged together yield a parameter which is the population _____.

=====
Mean

159. The technique for making this determination is called *analysis of variance*. Variance is the square of the standard deviation. The variance of a population having a standard deviation of 3 is _____.

=====
9

160. The means from several different groups could have a variance. This would be a measure of variation *between the groups* and is frequently called the *between group* variance. If the variation between the means of the groups were great, we would obtain a large _____ _____ variance. Thus, one takes the group means and derives a variance.

=====
Between Group

161. Each group has a standard deviation (and thus a variance) of its own. The mean of these variances would be a measure of the average variation *within the groups* and could be called the *within-group variance*. If the variation within a group were small, we would obtain a small _____ group variance. Thus, one takes the group variances and derives a mean of the variances.

162. Consider a crowd evenly divided between men and women. This crowd might have a mean height of 65 inches with a variance of 100 inches. This variance of the total sample could be called the *total* _____.

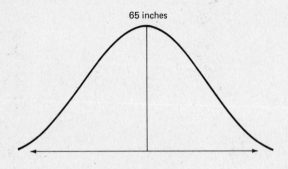

65 inches

163. Let us say that the men average 70 inches in height and the women 60 inches in height. Taken in total the statistics form the curve below. The variance of the curve is called the _____ variance. What is the curve's mean? _____.

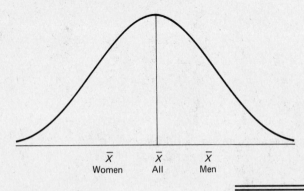

$\bar{X}$ Women $\bar{X}$ All $\bar{X}$ Men

60

164. Let's say that we now partition this crowd into 2 separate "cells" on the basis of sex.

men	women

The mean around which the men would cluster is now_____ inches. For women the mean is _____ inches.

$$\overline{\overline{}}$$

70   60

165. If the heights of the men alone vary less from the mean of men than from the mean of men and women together, the variance of men considered separately from women will have a smaller variance than the total variance.

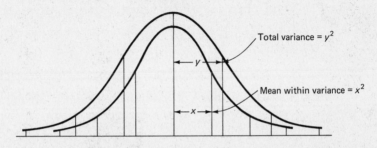

Total variance = y^2

Mean within variance = x^2

The case would be the same for women. The mean variance of the two groups individually would thus be _____ than the *total* variance.

$$\overline{\overline{}}$$

Smaller (or Less)

166. The means for each of these two groups are 60 inches and 70 inches, respectively. Since the groups cluster more closely about their new means, the two within-group variances _____.

$$\overline{\overline{}}$$

Shrink (or Decrease)

167. The greater the variation between groups as compared with the variation within groups, the greater will be the size of the _____ ratio.

$$\overline{\overline{}}$$

F

168. Pictured below are two different experiments, one by Dr. Jones and one by Dr. Smith. Both experiments compared the results of three different methods for teaching reading. In both experiments the means for the three conditions were 3, 4, and 5, respectively. The results within any one condition varied less in experiment A than B. Despite this, Dr. Smith's experiment showed significant results while Dr. Jones' did not. To see why, note that each of the groups in Dr. Smith's experiment was less variable (and thus his experiment had a smaller within variance) than those in Dr. Jones' experiment. Since the means were the same in each experiment, the F ratio (or the *between variance* divided by the *within variance*) would be larger in the experiment by Dr. _____.

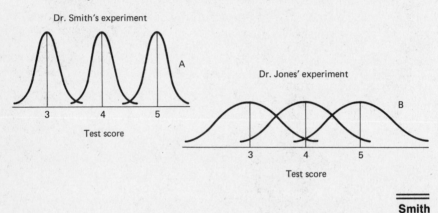

Dr. Smith's experiment

A

Dr. Jones' experiment

B

Test score

3 4 5

Test score

3 4 5

Smith

169. The basic question asked in analysis of variance is as follows: "Is the variation between groups enough greater than the variation within groups to reject any idea that the groups were samples of a common population?" Would a large F suggest an answer of yes or no? _____

Yes

170. The size of F required for significance varies by the number of groups, the number of people in each group, and the amount of confidence with which the investigator wishes to speak. Would the F required for the .01 level be higher or lower than that for the .05 level? _____

Higher

171. The sizes of *F* required for significance with particular sample sizes, numbers of groups, and confidence levels are to be found in the appendixes of most methods texts on statistics. (A brief example may be found in Table 3 following Frame 183.) These sources are appropriately called *F tables*. An *F* value of unity (1.00) or less is always nonsignificant; thus, if this were the case, it would not be necessary to consult an _____ table.

$$\frac{=}{F}$$

172. Based on the information we have just derived in the preceding frames, we can formulate an equation for the ratio between estimates of variability between groups and within groups.

$$F = \frac{\text{Estimate of variation between}}{\text{Estimate of variation within}}$$

In order to declare differences to be significant, a researcher would find the *F* ratio to be (high, low) _____.

$$\frac{\equiv}{\text{High}}$$

173. The measure of variability to be used in studying *F* ratios is called the *variance*. Since the decisions are based on estimates of variance between and estimates of variance within, the method is appropriately called Analysis of _____.

$$\frac{\equiv}{\text{Variance}}$$

174. Analysis of Variance (abbreviated ANOVA), while based on ratios of estimates of variance, is actually a tool for deciding the statistical significance of differences between _____.

$$\frac{\equiv}{\text{Means}}$$

Suppose that the visual perception tests (see Frame 123) had been given under different conditions to different children, the matching of child to condition being random. The psychologist might now ask if it is probable that any difference in means would appear between different segments of the population if they were to be given different treatments or, equivalently, whether the different treatments caused any difference between groups in the test scores other than the fluctuation that one would expect to occur by chance or, even more simply, whether the treatments really made a difference.

The assignment of treatments might have run something like: Don and Ray were warned of dire consequences should they do badly on the test; Jan, May, and Joy comprise a sample that received neither warning nor encouragement; Jim, Sam, Fay, and Art were told only that they would probably do very well.

The visual perception data, grouped by treatment, would now look like this.

Warned		Neutral		Encouraged	
Subjects	Scores	Subjects	Scores	Subjects	Scores
Don	0	Jan	2	Jim	4
Ray	4	May	4	Sam	4
		Joy	6	Fay	6
				Art	6

Even though it is generally desirable to have the number of subjects in each treatment group be as near to equal as is convenient, notice that equality is not necessary.

For the practice problem assume that Ima and Lil receive one treatment, while Dot, Hal, and Sue received the other. The breakdown would now be:

Treatment A		Treatment B	
Ima	0	Dot	2
Lil	2	Hal	3
		Sue	8

Don't try to fill in the answers for the next problem now. As you compute values in the next several pages, come back and fill in the blanks.

Notes	Example Problem				Practice Problem			
		SS	df	MS	F	SS	df	MS	F

175. As each value is calculated, fill in the blanks of the practice problem table. Correct entries will be given when the necessary calculations have been completed.

Example Problem:

	SS	df	MS	F
TOTAL	32	8		
BETWEEN	12/2	= 6.0		
WITHIN	20/6	= 3.3		$= 1.8$

Practice Problem:

	SS	df	MS	F
	__	__		
	__/__	= __	= __	= __
	__/__	= __		

Recall that the SS for the entire practice problem in Frame 104 (total SS) was:

$$\text{Total SS} = \Sigma X^2 - \frac{(\Sigma X)^2}{N}$$

$$= 0^2 + 2^2 + 2^2 + 3^2 + 8^2 - \frac{(0 + 2 + 2 + 3 + 8)^2}{5}$$

$$= 0 + 4 + 4 + 9 + 64 - \frac{15^2}{5}$$

$$= 81 - 45 = 36$$

Notes	Example Problem	Practice Problem

176. Calculate the SS for the entire sample as before (total SS). The degrees of freedom (df) for TOTAL = $N - 1$.

Example Problem:

$$\text{TOTAL SS} = \Sigma X^2 - \frac{(\Sigma X)^2}{N}$$

$$= 176 - \frac{36^2}{9} = 32$$

$$\text{TOTAL df} = N - 1 = 8$$

Practice Problem:

TOTAL SS = _____

TOTAL df = _____

——————
36   4

The total SS is composed of the SS between groups and the SS within groups. Since the total SS has been calculated, it is only necessary to calculate the SS between groups. The SS within groups can then be obtained by subtraction.

Notes	Example Problem	Practice Problem

177. Sum up all scores in a cell, square this total, and divide by the number of scores in the cell. Total up this value for all cells and then subtract CF. The result is the SS between groups (SS_{bg}).

$$SS_{bg} = \frac{(\Sigma X_1)^2}{n_1} + \frac{(\Sigma X_2)^2}{n_2} + \ldots - CF$$

$$\frac{(\Sigma X_1)^2}{n_1} = \frac{(0+4)^2}{2} = \frac{16}{2} = 8$$

$$\frac{(\Sigma X_2)^2}{n_3} = \frac{(2+4+6)^2}{3} = \frac{144}{3} = 48$$

$$\frac{(\Sigma X_3)^2}{n_3} = \frac{(4+4+6+6)^2}{4} = \frac{400}{4} = 100$$

$$SS_{bg} = 8 + 48 + 100 - 144$$
$$= 12$$

$SS_{bg} =$ _____

$\overline{\overline{13.33}}$

178. With k equal to the number of groups, the df between groups (df_{bg}) is equal to $k - 1$.

$$df_{bg} = k - 1$$
$$= 3 - 1$$
$$= 2$$

$df_{bg} =$ _____

$\overline{\overline{1}}$

179. The SS within groups (SS_{wg}) is equal to the total SS (SS_{Tot}) minus the SS between groups (SS_{bg}).

$$SS_{wg} = SS_{Tot} - SS_{bg}$$
$$= 32 - 12$$
$$= 20$$

$SS_{wg} =$ _____ $-$ _____

$=$ _____

$\overline{\overline{36 - 13.33 \qquad 22.67}}$

180. Similarly, the degrees of freedom within groups (df_{wg}) is equal to the total df (df_{Tot}) minus the df between groups (df_{bg}).

$$df_{wg} = df_{Tot} - df_{bg}$$
$$= 8 - 2$$
$$= 6$$

$df_{wg} =$ _____

$\overline{\overline{3}}$

66

For each of the components of the total variance, the estimate of variance or *mean square* (MS) is equal to the SS divided by the df.

Notes	Example Problem	Practice Problem

181. For MS_{bg} divide SS_{bg} by df_{bg}.

$MS_{bg} = SS_{bg}/df_{bg}$
$= 12/2$
$= 6$

$MS_{bg} =$ _____ / _____

$=$ _____

13.33/1   13.33

182. For MS_{wg} divide SS_{wg} by df_{wg}.

$MS_{wg} = SS_{wg}/df_{wg}$
$= 20/6$
$= 3.33$

$MS_{wg} =$ _____ / _____

$=$ _____

22.67/3   7.56

183. For F divide MS_{bg} by MS_{wg}.

$F = MS_{bg}/MS_{wg}$
$= 6/3.33$
$= 1.80$

$F =$ _____ / _____

$=$ _____

13.33/7.56   1.76

The completed ANOVA table for the practice problem should now look like:

	SS	df	MS	F
TOTAL	36.00	4		
BETWEEN	13.33	1	13.33	1.76
WITHIN	22.67	3	7.56	

The calculated F can now be compared with the values in a table showing F values that cut off the most extreme 5 percent of the chance distribution (see Table 3). To use this table merely follow across the top until the column for the df between groups is reached. For the example problem, $df_{bg} = 2$. Then follow the left-hand column down to the df within groups. For the example problem, $df_{wg} = 6$. The table value is 5.14.

Table 3   (For more complete tables, see pages 136 and 137.)

.05 Values for the Distribution of F
  Degrees of freedom

within groups	Degrees of freedom between groups		
	1	**2**	**3**
1	161.4	199.5	215.7
2	18.51	19.00	19.16
3	10.13	9.55	9.28
4	7.71	6.94	6.59
6	5.99	5.14	4.76
8	5.32	4.46	4.07
12	4.75	3.88	3.49
20	4.35	3.49	3.10
30	4.17	3.32	2.92
60	4.00	3.15	2.76
120	3.92	3.07	2.68
∞	3.84	2.99	2.60

184. In order for us to reject the null hypothesis (that no differences exist among the means) our calculated F would have to equal or exceed (meet or beat) this table value. Since our 1.80 does not meet or beat 5.14, we cannot reject the null hypothesis. We would report, "No significant difference (.05 level) was detected among the means."

    For the practice problem, our calculated F was _____, the df_{bg} was _____, df_{wg} was _____. The critical F would be _____.

    1.76   1   3   10.13
    ---

185. Since our calculated F (fails to, does) _____ meet or beat the table F, our decision follows that we (do not have evidence to, may) _____ reject the null hypothesis and conclude that (no evidence was found to suggest, it is most probable) _____ that the treatments make a difference in the population.

    **fails to   do not have evidence to
    no evidence was found to suggest**

When only two means were being compared, a technique called the *t* test has been much used. It involves just a bit more computation than analysis of variance and offers little in the way of advantages. If computation of a *t* between two independent means is called for, just compute *F* and take its square root. When only two groups are being compared, $t = \sqrt{F}$.

186. *F* tests and *t* tests deal with sample means and are incidentally based on the assumption that the distribution of the variables compared is pretty much normal (i.e., approximates a normal curve). Would it frequently be reasonable to assume that such values as achievement test scores, IQs, heights, weights, etc., would be normally distributed? _____ Would this assumption *always* be correct? _____

Yes	No

OPEN BOOK QUIZ

1. ANOVA is used primarily to test significance of differences between _____.

2. The Total SS is made up of the Between SS and the _____ SS.

   Assume an experiment with two conditions and two subjects per condition.

   CONDITIONS

b_1	b_2
1	5
3	7

Compute the:

a. Within SS _____
b. *F* _____

3. Is the difference in means between the two conditions significant at the .05 level? _____

69

Regression

187. Let us consider the problem of predicting human behavior. In our society this is a common need. Companies wish to predict which of their applicants would become good sales managers. Colleges and universities wish to select those applicants who would profit most if selected for study at their institutions. Prediction is a vital part of modern life.

   Let us suppose that we are trying to predict college freshmen grade point averages (henceforth called freshmen grades) from a paper and pencil test of academic aptitude taken during the senior year of high school (henceforth called the SAT Verbal). For our purposes, let us assume that the r between the SAT Verbal and freshmen grades is +.50. Thus, on the average, a person with a relatively high SAT Verbal would tend to earn relatively (high or low) _____ freshmen grades if he went to college with the same group to which he was compared when his relative standing on the SAT Verbal was determined.

High

188. There is an interesting, curious, and fundamental relationship between predictor values and the actions one tries to predict.

The predicted actions are more typical (on the average) than the actions used as predictors. Those who have very low values on the predictor will regress toward the mean future performance and thus improve. Those who do unusually well on the predictor also regress toward the mean future performance and thus average (better or worse) _____ in the future performance than on the predictor.

Worse

189. The principle is quite general. A group of men selected because they are very tall would tend to have sons who, when mature, would be closer to the mean of the general population and thus be shorter than the selected group. Likewise a group of men selected for shortness would tend to have sons who were (shorter or taller) _____ than themselves.

Taller

190. In the figure below, SAT Verbal scores and grades have been converted to z scores. The mean freshman grade of all those having an SAT Verbal score of $+2$ is $\dfrac{(+2) + (+1) + (0)}{3} =$

_____.

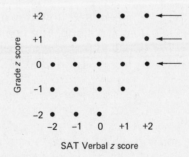

SAT Verbal z score

+1

191. Because the figure is symmetrical one might have supposed that the best match for a $+2$ on the SAT Verbal would be a $+2$

71

on freshman grades. Instead, the best estimate regresses closer to the mean. Fill in the missing values in the following table:

Actual SAT Verbal z score	Best Estimate for Freshman Grade z score
+2	+2 times .5 = +1
+1	+ _____
0	_____
−1	−1 times .5 = −.5
−2	_____

+0.5	**0.0**	**−1.0**

192. Connecting these points yields the following figure:

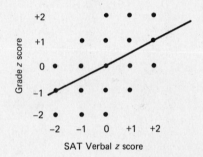

SAT Verbal z score

Notice that the line ascends .5 units for every unit it moves to the right. This is called a .5 slope. The line is called a regression line and its *slope* is called a *regression coefficient*. The figure above demonstrates a regression coefficient of_____.

$$\underline{\underline{.5}}$$

193. If one knew nothing about a possible future student, the best estimate of his freshman grade z score would be zero (i.e., the mean for all freshmen). Each SD of SAT Verbal allows us to venture another .5 SD away from a grade estimate of zero. What would be the best estimate of the freshman grade z score for an applicant with a SAT Verbal z score of +3? _____

$$\underline{\underline{1.5}}$$

72

194. The general regression formula, when *z* scores are used, is $\hat{y} = \beta'(z)$, where $\hat{y}$ (read "y hat") is the best estimate for the *z* score variable being predicted

β' (read "beta prime") is the regression coefficient based on *z* scores (and thus called the *normalized regression coefficient*) and

z is the *z* score of the predictor variable.

In the preceding frame, $\hat{y} =$ _____, $\beta' =$ _____, and $z =$ _____.

$$\overline{\quad\quad\quad\quad\quad\quad}$$
1.5   .5   3

195. It is interesting to note that the *normalized regression coefficient and r are identical.* Thus, if two tests had an *r* of .8, for each standard deviation an individual departed from the mean on one test, we could venture to predict that he would move _____ standard deviations away from the mean on the other.

$$\overline{\overline{\quad\quad\quad}}$$
.8

196. If the *r* between SAT Verbal and freshman grades were .4, one could best predict that an applicant having a SAT Verbal *z* score of −2 would have a freshman grade *z* score of_____.

$$\overline{\overline{\quad\quad\quad}}$$
−.8

197. Which of the scatter plots below reflects the highest normalized regression coefficient? _____ Which the highest *r*? _____

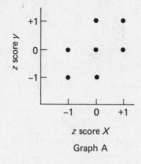

Graph A

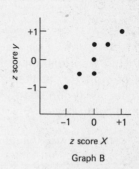

Graph B

$$\overline{\overline{\quad\quad\quad}}$$
B   B

73

198. Real data would seldom yield a perfectly straight line. The best regression line is mathematically determined so that the sum of the squares of the vertical deviations of the actual data from the line is at a minimum. Which of the lines below would be the best by this criterion, A or B?

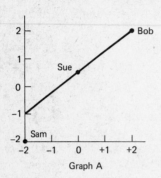

Graph A

	Deviations	Deviations Squared
Bob	0	0
Sue	0	0
Sam	−1	1
		Total 1

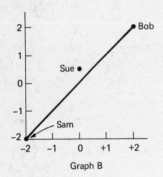

Graph B

Bob	0	0
Sue	$\frac{1}{2}$	$\frac{1}{4}$
Sam	0	0
		Total $\frac{1}{4}$

B

199. If a train were 650 miles from a station at zero time (say noon) and going away at the rate of 30 miles an hour, one could tell how far away the train would be at any time by using the following formula:

650 + 30 times the number of hours after noon

For example, at 2:00 P.M. the train would be 650 + 30 (2) = _____ miles away.

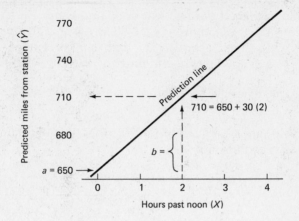

$$\overline{\overline{710}}$$

200. If the predicted distance is $\hat{Y}$, the point at which the line cuts the ordinate (650 in this case) is a, the slope (30 in this case) is b, and the predictor variable (time in this case) is X, the general prediction formula would be $\hat{Y} =$ _____.

$$\overline{\overline{a + bX}}$$

201. In day-to-day practice, regression values are usually expressed in the most available units. Assuming, for convenience, a freshman grade mean of 80 for all students with an SD of 6, the figure might look like this:

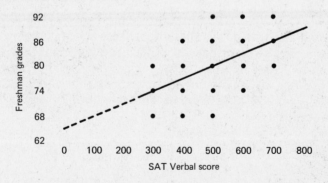

What would be the best grade estimate for an applicant with an SAT Verbal of 700? _____

$$\overline{\overline{86}}$$

75

202. Since predicted grades start at 65 and climb 3 grade points for every 100 SAT points (i.e., 0.03 grade points for every SAT point), the best estimate for high school grades in this particular example may be obtained by the following formula:

65 + 0.03 times the SAT score

If a student's SAT score is 200, what specific numbers would go with each symbol?

$\hat{Y} = $ _____ $a = $ _____ $b = $ _____ $X = $ _____

71.00	65.00	00.03	200.00

203. Note that the value of the ordinate (vertical line) is 65 where the regression line cuts the abscissa (the horizontal axis) 0. From that point, grades soar 3 points for each rise of 100 SAT points (i.e., .03 to 1). One could thus predict any particular case by the following formula:

Predicted grade = 65 + .03 (verbal SAT)

The predicted grade for an applicant with an SAT of 600 is 65 + .03 (600) = 65 + 18 = 83. What is it for an SAT Verbal of 500? _____

80

204. Just as *r* was called *linear* because a straight line was involved, the regression discussed in this program is called *linear regression*. Prediction is usually improved by the fairly complex process of simultaneously using several variables to predict one variable. The technique is called multiple _____ regression.

Linear

205. The regression line has a formula which, if known and applied, can be used with varying success to predict anything, once a predictor's relationship to the predicted is known. The equa-

tion of the line using raw data involves a constant plus the _____ coefficient times an individual score.

$$\overline{\overline{\text{Regression}}}$$

206. As an example of how a regression analysis might be used to predict behavior, suppose that an admissions director for a college wanted to prophesy the performance of students in their freshman year. He would first get, for example, the SAT scores of a group of students (the X scores) and follow up by getting the grade point averages for that same group (the Y scores). From these pairs of X and Y scores he could figure out values for a and b. Now let us suppose that a student applies to that college. If the director had that student's SAT score (X), he could get a prediction or prophesy of that student's grade point average ($\hat{Y}$) by the formula _____.

$$\overline{\overline{a + bX}}$$

Predicting Behavior: Practical Computations

The values a and b for actual data are easily calculated.

Notes	Example Problem		Practice Problem	
207. From the Section on r (Frames 124 to 127) recall that:	$M_x = 4$	$M_y = 5$	$M_x = $ _____	$M_y = $ _____
	$SS_x = 32$	$SS_y = 50$	$SS_x = $ _____	$SS_y = $ _____
	$SP = 24$		$SP = $ _____	

$$\overline{\overline{\begin{array}{cc} 3 & 2 \\ 36 & 4 \\ & -7 \end{array}}}$$

208. To compute b, divide SP by SS_x	$b = SP/SS_x$	$b = $ _____ $/$ _____
	$= 24/32$	
	$= .75$	$= $ _____

$$\overline{\overline{\begin{array}{c} -7/36 \\ -.194 \end{array}}}$$

209. To compute a, subtract b times the mean of X from the mean of Y

$a = M_y - bM_x$

$= 5 - (.75 \times 4)$

$= 2$

$a = \underline{\hspace{1cm}} - (\underline{\hspace{1cm}} \times \underline{\hspace{1cm}})$

$= \underline{\hspace{1cm}}$

$2 - (-.194 \times 3)$
2.58

Let us assume a predictor (X) score of 4 for the individual involved in both the example and the practice problem.

Notes	Practice Problem	Example Problem

210. To predict Y (Predicted Y is symbolized here as $\hat{Y}$) for a particular individual, add b times the X score for that particular individual to a.

Practice Problem:

$\hat{Y} = a + bX$

$= 2 + (.75 \times 4)$

$= 2 + 3$

$= 5$

Example Problem:

$\hat{Y} = \underline{\hspace{1cm}} + (\underline{\hspace{1cm}} \times \underline{\hspace{1cm}})$

$= \underline{\hspace{1cm}} + \underline{\hspace{1cm}}$

$= \underline{\hspace{1cm}}$

$2.582 + (-.194 \times 4)$
$2.582 + (-.776)$
1.806

In summary, consider what we have just done in the example problem. Having gathered data from a sample of 5 subjects, a regression equation was derived. It was $\hat{Y} = 2.97 - .194 X$. Now along comes a subject with an X score of 4. We were able to use the prediction equation and prophesy that his Y score would be 2.19.

1. If the correlation between two variables was .6, a rise of 2 z score points in the X variable would cause one to predict a rise in the Y variable of _____ a score points.

     Consider the following set of scores:

X	Y
0	2
2	4
4	3

2. Calculate b. _____

3. Calculate a. _____
4. Predict Y (i.e., compute $\hat{Y}$) if $X = 4$. _____
5. For which of the following scatter diagrams would a regression analysis (as we have studied it) be most appropriate? _____

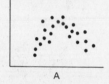
A

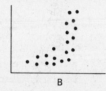
B

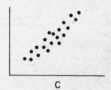
C

Answers	Comments
1. 1.2	$= r = .6$ $y = (z)$ $y = .6 (2) = 1.2$
2. .25	$SP = 20 - (6 \times 9/3) = 2$ $SS = 20 - (6^2/3) = 8$ $b_x = 2/8 = .25$
3. 2.5	$a = M_y - b M_x$ $= 3 - .25 (2)$ $= 3 - .5 = 2.5$
4. 3.5	$Y = a + bx$ $= 2.5 + .25 (4)$ $= 2.5 + 1 = 3.5$
5. C	The type of regression analysis we've studied is only appropriate for *linear* (straight line) relationships.

Chi-Square (Finding Out if Frequencies Are Different than Expected)

Analysis of variance deals with means, but sometimes a psychologist wishes to study frequencies.

211. Consider the question: Are more babies born in one season than another? To investigate this, we could start off by gathering some birth dates from a randomly selected sample. Let's just suppose that we now have a total of 40 birth dates, and that we have defined the seasons so that spring, summer, fall, and winter each has an equal number of days. How many births out of the 40 would you *expect* in each season if, in fact, season did not make a difference?

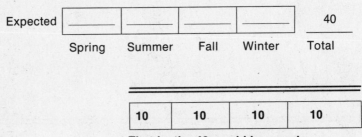

Expected					40
Spring	Summer	Fall	Winter		Total

10	10	10	10

That is, the 40 would be evenly distributed into each season.

212. Now it might turn out that when we actually grouped the *observed* birthdates of the 40 person they were distributed thus:

Observed	15	8	5	12
	Spring	Summer	Fall	Winter

The question now would be: Is the fact that the *observed* are different from what we *expected* more likely due to chance or does it more likely represent actual population differences in birth rate during the different seasons?
To decide this question of whether *frequencies* are sig-

nificantly different than expected, a very popular, easy, and useful test is the *Chi-square test.*

To perform this test: 1. Summarize the information so far as below, with E being in the small box for each cell.

E = 10	E = __	E = __	E = __
O = 15	O = __	O = __	O = __

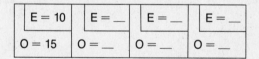

213. Separately, for each cell, subtract the *expected frequency* (E) from the *observed frequency* (O).

| O − E | 15 − 10 = 5 | _____ | _____ | _____ |

|−2|−5|2|

214. Square these differences.

| (O − E)² | 25 | _____ | _____ | _____ |

|4|25|4|

215. Divide each of these squared differences by E.

| $\frac{(O - E)^2}{E}$ | $\frac{25}{10} = 2.5$ | _____ | _____ | _____ |

|.4|2.5|.4|

81

216. Add all of these up and the result is called *chi-square* (χ^2).

$$\text{Chi-square} = \sum \frac{(O - E)^2}{E} = 2.5 + \underline{\hspace{1cm}} + \underline{\hspace{1cm}} + \underline{\hspace{1cm}} = \underline{\hspace{1cm}}$$

$$.4 + 2.5 + .4 = 5.8$$

217. A relatively large chi-square should indicate that the E's differed more from the O's than is likely by chance. As to how large a value for chi-square is needed to reject the null hypothesis of no difference in the population, we again consult a table (Table 4). To enter the table, find the degrees of freedom by using the number of cells minus one. In the present case, we have 4 cells (the four seasons). The degrees of freedom will be _____.

$$4 - 1 = 3$$

Table 4   (For a more complete table, see page 139.)

.05 Points for the Distribution of Chi-Square

Degrees of freedom	Chi-square
1	3.84
2	5.99
3	7.81
4	9.49
6	12.59
8	15.51
12	21.03
20	31.41
30	43.77
60	79.08
120	146.57

218. For this problem, what value of chi-square would we have to meet or beat in order to declare that a significant (.05 level) dif-

ference exists between what we would *expect* if there was no population difference and what we have *observed*? _____

$\overline{\overline{\text{7.81}}}$

219. Since our computed χ^2 of 5.80 fails to meet or beat 7.81, we may now (reject, fail to reject) _____ the hypothesis that only chance has caused any differences between *expected* frequencies and *observed* frequencies.

Fail to Reject

   To put our finding another way, we would conclude that, while there might be a difference in birth rate between the various seasons, our data do not provide enough evidence to bet in that direction. This is not the same as saying there is no difference. Consider the way that many people feel about elves (of the wee, fey variety). These people find no evidence that elves exist but, on the other hand, they have failed to prove the nonexistence of elves. Even though you've peered under a million mushrooms, an elf could be sitting beneath the very next mushroom.

Two-Dimensional Chi-Square

220. Chi-square analysis can also be used to find out if there is a relationship between how people fall into various categories. Chi-square analysis can show whether there is a probable relationship between voting pattern and ethnic group, between cars driven and income level, sex and color preference, educational level and type of music preferred, etc.

   As an example, assume a six-room house where only men were allowed in one side and only women in the other. Different drinks were served in each set of rooms so that the house plan would look like the figure below.

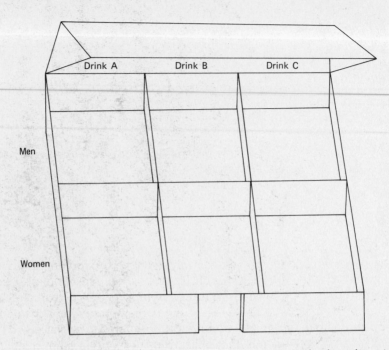

A large party is now held in the house and when it gets going well, we gently lift the roof off the house and, peering inside, we see what is illustrated in the figure below.

Observe the frequency in each category (room).
What is the *observed frequency* (O) in each category?

Drink A   Drink B   Drink C

	Drink A	Drink B	Drink C
Men			
Women			

	Drink A	Drink B	Drink C
Men	10	5	15
Women	5	10	5

221. And now to find out if there is some relationship between sex
and drink preferred? It looks like women might tend toward
drink B and men toward drink C, but chance could cause some
effect in our sample even if there was no relationship in the
population.

   It's the null hypothesis approach again. We first hypothe-
size that there is no relationship between sex and drink preferred.

   If there were no relationship, given that 30 out of the 50
people are men (60 percent), what percent of those preferring
drink C would you *expect* to be men? _____

60 Percent

222. Since 20 persons preferred drink C and you would expect 60
percent to be men, what would be the expected frequency (E)
for men preferring drink C? _____

60 Percent of 20 = 12

85

223. We can summarize the separate effects by totaling around the margins. These totals are called *marginals*.
   Complete the marginals for this table.

	Drink A	Drink B	Drink C	
Men	10	5	15	30
Women	5	10	5	___
	___	___	___	

$$\overline{}\ 20$$

15   15   20

---

224. The sum of the expected frequencies in one column (or line) also equals the marginal.
   Since 20 persons prefer drink C and the E for men is 12, the E for women for drink C must be _____.

$$\overline{}$$
$$20 - 12 = 8$$

---

225. In case this intuitive approach to getting E's isn't clear to you, an alternate formula may help.

$$E \text{ for any square} = \frac{(\text{Column marginal}) \ (\text{Row marginal})}{\text{Grand total for all cells}}$$

For the present example,

$$E \text{ for men, drink B} = \frac{15 \times 30}{50} = \frac{450}{50} = 9$$

Compute the E for the remaining cells and complete this table.

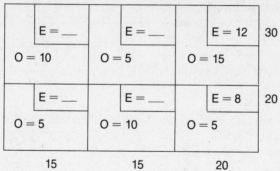

E = ___ O = 10	E = ___ O = 5	E = 12 O = 15	30
E = ___ O = 5	E = ___ O = 10	E = 8 O = 5	20
15	15	20	

| 9 | 9 | 12 |
| 6 | 6 | 8 |

226. With O and E for each cell in hand, the final calculations are easy.

For each separate cell, compute $(O - E)^2/E$.

| $(O - E)^2/E = (10 - 9)^2/9$
$= -1^2/9$
$= 0.111$ | _____ | _____ |
| _____ | _____ | _____ |

	Drink A	Drink B	Drink C
Men	0.111	1.778	0.750
Women	0.167	2.667	1.125

87

227. The sum of $(O - E)^2/E$ for all of these cells is called *chi-square*. For the present problem chi-square $(\chi^2) = 0.111 + 1.778 + 0.75 + 0.167 +$ _____ $+$ _____ $=$ _____.

228. If the observed frequencies were quite different from the expected, we would find each $(O - E)^2$ and thus $\Sigma(O - E)^2/E$ to be relatively (small, large) _____.

229. To enter the chi-square table when doing a two-dimensional chi-square analysis, find the degrees of freedom by multiplying the number of rows minus 1 by the number of columns minus 1. In this case it is $(2 - 1)(3 - 1) = 2$.

    In the present problem, what value of chi-square would we have to meet or beat in order to declare a significant (.05 level) relationship between sex and preferred drink? _____

230. Since our computed χ^2 of 6.598 beats the critical value of 5.99, at the .05 level, we may now (reject, fail to reject) _____ the null hypothesis of no relationship.

    Two final points on chi-square: If there is only one degree of freedom, a special adjustment (subtracting 0.5 from the absolute value of $O - E$ before squaring) is generally applied.

    Chi-square probabilities are not very accurate unless at least 80 percent of the E's are greater than 5.

1. To find out if means differ significantly from each other, use
_____ _____ _____ but to find out if
frequencies differ from the expected, use _____.

2. A sample of 30 voters was asked which of 3 candidates they
planned to vote for with the following results:

Test the hypothesis that, in
the population, there is no
difference in preference for
any of the candidates. Conclusion? _____

Candidate	Frequency
Smith	5
Jones	20
Baker	5

3. A sample of husbands and wives were asked their preference on
car size and the results came out thus:

Husbands	20	30	10
Wives	10	20	30

Small   Medium   Full

If there is no difference between the preference pattern of husbands and wives in the population, what frequency would you
expect for husbands preferring small cars? _____

4. For Problem 3, compute chi-square _____.

5. For Problem 3, what chi-square would be needed for significance
at the .05 level? _____

Answers	Comments
1. Analysis of Variance (or ANOVA)	See Frames 158 and 211
2. There is a significant difference Chi-square	Expected is $\boxed{10}$ $\boxed{10}$ $\boxed{10}$ $\chi^2 = (5-10)^2/10 + (20-10)^2/10 + (5-10)^2/10$ $= 2.5 + 10 + 2.5 = 15$ At 2 df, χ^2 needed is 5.99
3. 15	$(60 \times 30)/120$
4. 15.34	$1.67 + 1.67 + 1 + 1 + 5 + 5$ 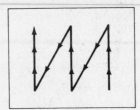
5. 5.99	$df = (\text{Rows} - 1)(\text{Cols.} - 1) = 1 \times 2$

Confidence Intervals

231. Up to this point inferential methods have been used to: (1) make point estimates of parameters (as means and standard deviations; (2) test the significance of correlation coefficients; (3) predict behavior (regression analysis); and, (4) conduct tests for the significance of differences between means (ANOVA) and be-

tween expected and observed frequencies (chi-square). The next example represents a class of inferential methods that yield upper and lower boundaries for intervals within which — one can feel rather confident — a parameter lies.

If we wished to know the mean SAT score for a college and had only the resources to test a random sample of students, we would find (as before) that the best estimate was 525 (for example). We could, however, additionally show that there was a 95 percent chance that the population mean was in the interval from 514 to 536 (for example).

To get some insight into this approach, consider the cases in the figure below as a population. Let us now take a random sample of 4 scores, compute their mean, and plot it on the X axis.

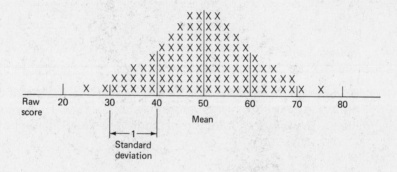

232. If the first sample of scores was 40, 50, 70, and 80, place an X where its mean would be on the line below:

20 30 40 50 60 70 80

```
                              X
20   30   40   50   60   70   80
```

233. In similar manner, compute the mean and place an X on the line below for the means from each of these successive samples:

Sample number	Hypothetical values	Mean
1st	40, 50, 70, 80	60
2nd	40, 50, 55, 55	—
3rd	30, 35, 35, 80	—
4th	45, 45, 50, 60	—
5th	30, 35, 40, 55	—

```
                        X
          _____
          20  30  40  50  60  70  80
```

```
                                    60
                                    50
                                    45
                                    50
                                    40

                                    X
                                 XXX   X
          _____
          20   30   40   50   60   70   80
```

234. Since extreme values would tend to cancel out, the distribution of means should be (less, more) _____ variable than the original distribution of single scores.

Less

235. The standard deviation of these means, called the *standard error of the mean* (symbolized s_m), is easily estimated from the formula

$$s_m = \frac{s}{\sqrt{N}}$$

where s is the estimated standard deviation and N is the sample size. This relationship pops up in so many formulas and concepts that it may be to the behavioral scientist what $E = mc^2$ is to the physicists.

   In the present example, if s = 10 and N = 4,

$s_m = s/\sqrt{N} = $ _____ $/\sqrt{\text{_____}} = $ _____

$10/\sqrt{4} = 5$

236. Compute s_m for the visual perception scores given earlier in Frame 101. _____

Recall that these 9 scores had an s of 2 (Frame 144).

$$s_m = s/\sqrt{N} = 2/\sqrt{9} = 2/3 = .67$$

Distribution of Means

237. No matter what the shape of the original distribution (skewed, bimodal, rectangular, normal, etc.), a wondrous event occurs with the distribution of the means. It approaches the normal distribution. Since the mean of all the means would naturally approach the population mean, the theoretical distribution of means can be completely described.

With sample sizes equal to 4, the population mean equal to 50, and the population standard deviation equal to 10, the standard error of the means would equal _____, the mean of the means would equal _____, and the distribution would be close to _____.

$5(\text{i.e., } 10/\sqrt{4})$	50	Normal

Since the means are normally distributed, it follows that about 95 percent of the means would be within 2 standard errors of the population mean.

The experimenter does not, of course, have a large number of means. He has only his sample mean, and his need is not so much to know how sample means distribute, but how to use the values he has (M, s, and N) to get an idea of how far away the population mean is likely to be from his particular sample mean.

238. The following medieval spy story may help illustrate the logic behind how this is done. There was once a walled village named Gaussburg. In the very center of the village was a famous mint.

Map of
Gaussburg

|←——— 2 Miles ———→|←——— 2 Miles ———→|

The diagram is a map of the village and its 100 houses. The large building marked M is the mint. Notice that about $2\frac{1}{2}$ percent of the houses are beyond 2 miles upstream from the mint and about $2\frac{1}{2}$ percent are beyond 2 miles downstream.

A roving band of numismatists from Würm wished to position a spy within 2 miles of the mint. Their problem was to determine how likely it was that the mint would be within 2 miles of any house chosen at random from the classified shouts of the town crier.

The map shows that 95 percent of the houses were within 2 miles of the mint.

The entire band was puzzled about how to use this information until it was realized that, conversely, the mint itself must be within 2 miles of _____ percent of the houses.

$$=\atop{95}$$

Feeling 95 percent confident now that the mint would be within 2 miles of any house they might randomly acquire, the numismatists finally had peace of mind. This period later came to be known as the "peace of the Würm mint spy."

239. In other words, 95 percent of sample means (given that the sample size is moderately large) will fall within about 2 standard errors of the population mean.

You may feel 95 percent confident that, for any particular

sample mean you might gather, the population mean is not more than _____ standard errors away.

$$= \frac{}{2}$$

240. With small samples (below 20 or so) s (and thus s_m) starts to become erratic, so one has to cast the net somewhat further than $2s_m$ on each side of the sample mean in order to have some particular surety (as 95 percent) of capturing the population mean.

In this circumstance one computes s_m just as before, but, instead of multiplying by 2, the s_m is multiplied by a special value (t) which depends on the sample size and the probability with which one wishes to be sure of catching the population mean. For example, if one wishes to set boundaries that have a 95 percent chance of including the population mean, he could use the t table (Table 5).

Table 5 (For a more complete table, see page 138.)

.05 Values for the Distribution of t

$N-1$	t
1	12.71
2	4.30
3	3.18
4	2.78
6	2.45
8	2.31
12	2.18
20	2.09
30	2.04
60	2.00
120	1.98
∞	1.96

In the case of the visual perception scores, since N = 9, the t value is 2.31 (i.e., using Table 5, N − 1 = 9 − 1 = 8, and the t value for df = 8 is 2.31). Multiplying the standard error of the mean (.67) by t (2.31) yields 1.55. The mean of the population has a 95 percent probability of being within 1.55 of the sample mean. Since the sample mean is 4, the population mean has a 95 percent probability of being between 2.45 (4 − 1.55) and 5.55 (4 + 1.55). This interval from 2.45 to 5.55 is called the 95 percent confidence interval (see figure on page 96).

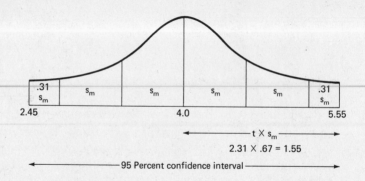

.31 s_m 2.45 s_m s_m s_m s_m 4.0   .31 s_m 5.55

t × s_m
2.31 × .67 = 1.55

95 Percent confidence interval

If for a sample of N = 64 the s is 16 and the mean 10, an estimate of the standard error of the mean would be _____ and the nearest *t* for the 95 percent would be _____. Their product is _____. The confidence interval thus extends from a lower confidence limit of 10 − _____ = _____ to an upper confidence limit of 10 + _____ = _____.

2	2.0	4	10 − 4 = 6	10 + 4 = 14

OPEN BOOK QUIZ

Assume that a sample of 64 scores had an M of 90 and an s of 4:

1. Estimate the standard error of the mean. _____
2. What would be the appropriate value of *t* to use in setting up a 95% confidence interval? _____
3. How many points away from the sample mean would the boundaries of the 95% confidence interval lie? _____
4. Compute the lower and upper boundaries for the 95% confidence interval. Lower boundary (or limit) = _____. Upper boundary = _____.
5. What is the probability that the actual population mean lies outside the 95% confidence interval? _____

Answers	Comments
1. 0.5	$s_m = s/\sqrt{N} = 4.0/\sqrt{64} = 4.0/8.0 = .5$
2. 2.00	$N - 1 = 64 - 1 = 63$ is not available in the t table, but 60 is very close. If $N - 1$ was much more diverse from any given table value, one could interpolate or, better yet, locate a more complete table.
3. 1.00	t times $s_m = 2.00$ times $0.5 = 1.00$
4. 89.0; 91.0	Lower boundary = $M - (t \times s_m) = 90.0 - 1.0$ Upper boundary = $M + (t \times s_m) = 90.0 + 1.0$
5. .05 or 5%	If the probability that the true or actual population mean lies within the confidence interval = 95%, then 100% − 95% = 5%

Interpreting Test Scores

241. The number of correct answers that a person acquires on a test is called his raw score. Assuming that each question on a test counted one point, a raw score of 12 would mean that an individual answered _____ questions correctly.

12

242. The raw score alone is a poor indicator of test performance. For example, a raw score of 40 points out of a possible total of 40 points suggests good performance. A raw score of 40 points out of a possible 400 suggests poor performance. In both cases 40 points is the _____ _____.

Raw Score

243. A somewhat better indicator of test performance is the percentage score (number right/total number of questions × 100). What percentage score does 100 right out of 400 questions receive? _____

25 Percent

244. Raw scores or percentage scores do not signify a student's performance *relative to the rest of the group.* A person cannot tell whether a percentage score of 91 percent is good or bad unless he knows the other students' scores. It is possible that all the other students received scores higher than 91 percent. In such a case 91 percent would be a low score _____ to the rest of the group.

Compared (or Relative)

245. If scores are ranked from lowest to highest, we can indicate a person's position relative to the group by stating, for instance, that out of a group of ten, he ranks fourth *from the bottom.* What is the rank of the person with a test score of 91 percent in the distribution 83 percent, 85 percent, 88 percent, 91 percent, 96 percent? _____

Fourth (from the bottom)

246. While the sizes of the two hypothetical groups in the preceding frame were different (10 and 5), the selected persons had the same _____.

Rank

247. The person who ranked fourth in a group of ten was not as relatively high as the person who ranked fourth out of a group of five. A rank by itself has little meaning unless the *total number* of persons in the group is known. The weakness of ranks then is that they depend upon the _____ of persons in a group.

Number (or Total Number)

248. The disadvantages of raw scores, percentage scores, and ranks are avoided by converting a person's rank to the number of cases he would equal or surpass if there were 100 cases in the group. Groups of different sizes are more directly com-

parable by saying what a person's rank would be out of a group of _____ .

$$\overline{\overline{100}}$$

249. If there were 5 persons ranked 1 through 5, each person would represent $\frac{1}{5}$ or 20 percent of the total group. The person with the rank of 4 would represent _____ percent of the group.

$$\overline{\overline{20}}$$

250. The 3 persons ranking below the person with the rank of 4 would represent $\frac{3}{5}$ or _____ percent of the group.

$$\overline{\overline{60}}$$

251. If the person (or score) with the rank of 4 is thought of as in the middle of the percentage he represents, half of the 20 percent (or 10 percent) of the total group could be thought of as having the same rank but as exceeded by the person (or score) with the rank of 4. The person (or score) with the rank of 4 (out of 5) thus exceeds: 60 percent by virtue of ranking above them and an additional 10 percent by virtue of being considered in the middle of those ranking the same. The total percent of scores it exceeds is then _____ percent.

$$\overline{\overline{70}}$$

252. If there were 10 scores ranked: 1, 2, 3, 4, 5, 6, 7, 8, 9, 10, the score ranking 8 would certainly exceed all scores ranking below it which are _____ percent of the total group.

$$\frac{}{70}$$

253. Since the score ranking 8 may, in addition, be thought of as exceeding half of the percentage it represents (i.e., ranking the same), it exceeds an additional half of 10 percent (or _____ percent). The total percentage it exceeds is thus _____ percent + _____ percent = _____ percent.

$$\frac{}{5 \qquad 70 + 5 = 75}$$

254. The Percentile Rank (PR) of a score is most often defined as the percentage of scores having a lower rank plus one half the percentage of scores having the same rank. The score that ranked 6th out of 25 would have a PR = _____ +
% ranks below

_____ percent = _____.
$\frac{1}{2}$ the % this rank

$$\frac{}{20 + 2 = 22}$$

255. A formula that may be convenient to use is:

$$PR = \frac{(N \text{ below the score}) + .5 (N \text{ at the score})}{Total\ N\ of\ the\ group} \cdot 100$$

Thus if one score ranked 6th out of 25,

$$PR = \frac{5 + .5(1)}{25} \cdot 100 = \text{_____}.$$

$$\frac{}{22}$$

256. Compute the PR of the person with the test score of 91 in the distribution 83, 85, 91, 92, 95, 95, 95, 96, 97, 98.

$$\frac{}{25}$$

100

257. What is the PR of Joe's score? Jan 82, May 81, Ray 85, Joe 87.

$$\overline{\overline{87.5}}$$

258. What is the PR of Bob's score? Ann 77, Bob 81, Don 90, Nan 80, Ned 87, Pat 75, Sue 92. Don't forget to rank the scores.

$$\overline{\overline{50}}$$

259. In the distribution 20, 20, 20, 30, 40, 40, 60 the ranks of 30 and 60 are obviously 4 and 7, respectively, but the ranks of the tied scores 20 and 40 are less obvious. The ranks occupied by 20 are 1, 2 and 3. The ranks occupied by 40 are _____ and _____ .

$$\overline{\overline{5 \qquad 6}}$$

260. When scores of the same value occupy more than one rank, the mean of those ranks is regarded as the tied-scores rank. For example, the rank of 20, that is, the mean of the ranks 1, 2, and 3, equals 2. The tied score 40, which occupies ranks 5 and 6, has a rank of _____ .

$$\overline{\overline{5.5}}$$

261. The ranks for the whole distribution now run as follows:

Raw score:	20,	20,	20,	30,	40,	40,	60
Rank:	2,	2,	2,	4,	5.5,	5.5,	7

Did the calculations for tied ranks affect the ranks of untied scores? _____ .

$$\overline{\overline{No}}$$

262. In the distribution 19, 22, 26, 26, 26, 34, 35 the tied scores of 26 have a rank of 4 and the scores of 34 and 35 have the ranks of _____ and _____ , respectively.

$$\overline{\overline{6 \qquad 7}}$$

101

263. In the distribution: 20, 20, 20, 30, 40, 40, 60, what is the PR of 40?

$$\frac{4 + .5(2)}{7} \cdot 100 \text{ or } \tfrac{5}{7} \cdot 100 \text{ or } 71 \text{ (rounded)}$$

Standard Score

264. Since raw scores tend to cluster around the mean of a distribution the percentile ranks do not have a constant relationship to the raw scores. From what is present of the following percentile-rank table, would it be possible to predict the omitted raw score? _____

Raw score	Percentile rank
10	5
22	10
23	15
28	20
?	25

No

265. Because percentile ranks do not have a constant relation to their scores it is meaningless to average percentile ranks. Could you properly average the following percentile ranks? 2, 3, 5, 10? _____

No

266. A measure of relative standing that has a constant relationship to raw scores and that can be legitimately averaged is the *standard score*. Standard scores can be computed for any distribution but their interpretation is difficult if the distribution is not normal. The conclusions of this program are valid only when the

distribution is near normal or normal. From what is present of the following table, predict the omitted raw score. _____

Raw score	Standard score
40	−2
45	−1
50	0
55	+1
?	+2

$$\equiv$$
60

z Score

267. A standard score is expressed in units of the standard deviation. It tells us how many standard deviations a score is above or below the mean. A score 2 standard deviations above the mean would have a standard score of +2 while a score 1 standard deviation *below* the mean would have a standard score of

_____.

$$\equiv$$
−1

268. As previously noted, this type of standard score is called a z score. What is the z score of a person who ranks 3 standard deviations above the mean? _____

$$\equiv$$
+3

269. The z score can be conveniently computed by dividing a score's deviation from the mean by the standard deviation $z = x/\sigma$. If the mean of a distribution is 70 and the standard deviation is 6, what z score is assigned to the person obtaining a score of 82?

_____.

$$z = \frac{82 - 70}{6} = +2$$

103

270. The mean raw score has a z score of _____ and (assuming a symmetrical distribution) would have _____ percent of the cases lying below it.

271. While z scores are fine, they do have two awkward characteristics: the plus or minus sign and decimal points. A way around these problems is to multiply the z score by some convenient constant (thus eliminating the need for decimal points) and then to add some constant that makes all of the scores positive. For example, if a z score of −2.00 was multiplied by 100 we would have _____. If 500 was added to this, the new score would be _____.

272. In just this way, the mean for College Board (CEEB) scores (SAT) is arbitrarily set at 500 and the standard deviation at 100. An SAT score of 400 indicates that a person is 1 standard deviation below the mean. (See the figure on page 105.)

An SAT score of 700 would indicate that a person had a raw score _____ standard deviations _____ the mean.

Normal distribution. A crowd of 1000 persons viewed from an airplane might look like this if each person lined up behind a sign giving his standard score on some text. Caution: Each type of standard score is in reference to the persons who took that particular test, so that one type of score can not be compared directly to another. For example, an IQ score gives one's standing compared with the entire population, while an SAT score gives one's standing compared with only those who took the SAT test.

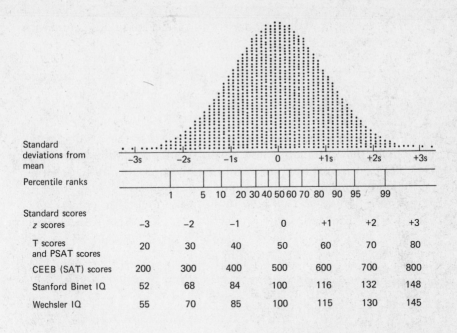

Standard deviations from mean	−3s	−2s	−1s	0	+1s	+2s	+3s
Percentile ranks		1	5 10 20 30 40 50 60 70 80 90 95 99				
Standard scores z scores	−3	−2	−1	0	+1	+2	+3
T scores and PSAT scores	20	30	40	50	60	70	80
CEEB (SAT) scores	200	300	400	500	600	700	800
Stanford Binet IQ	52	68	84	100	116	132	148
Wechsler IQ	55	70	85	100	115	130	145

273. One application of the standard deviation lies in the modern derivation of the Standard Binet IQ scores which have a mean of 100 and a standard deviation of 16 (see figure above). A person who is 1 standard deviation above his age group on the raw test score would be assigned an IQ score of 116 (100 + 16), 2 standard deviations above would get an IQ score of 132 [100 + (2 × 16)], $\frac{1}{2}$ standard deviation below the mean would yield an IQ of 92 [100 + (−.5 × 16)], etc. How many standard deviations above the mean of his age group is a person with an IQ of 124?

———

$$\equiv$$
1.5

274. Convert each of the following z scores into a standard score by multiplying each score by 100 and then adding 500.

z score	Standard score
−3	_____
−1	_____
0	_____
+.5	_____

$$\equiv$$
200
400
500
550

275. SAT scores and the modern interpretation of IQ scores are examples of _____ scores.

276. Since standard scores are based upon the model of the normal curve, they are inept when describing a population whose distribution is not normal. Would standard scores aptly describe a person's position in a bimodal population? _____ Would percentile ranks be more appropriate? _____

OPEN BOOK QUIZ

1. Compute the percentile rank of the raw score "8" in the following distribution: 8, 5, 6, 9, 3, 5. _____

2. Which (of percentile rank or some standard score) would be the most practical data for research where one was averaging, comparing, etc.? _____   _____

3. If the mean raw score for a test was 75 with a SD of 5, what would be the z score for Joe if he had a raw score of 83? _____

4. Convert a z score of $+2$ into a standard score with a mean of 50 and an SD of 10. _____

5. Assuming that IQ scores approach being normally distributed, about what percent of IQ scores are below 84? _____

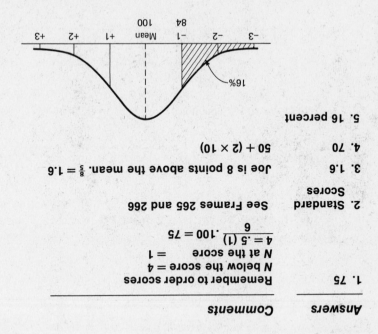

Answers	Comments
1. 75	Remember to order scores N below the score = 4 N at the score = 1 $4 = \frac{.5\ (1)}{6} \cdot 100 = 75$
2. Standard Scores	See Frames 265 and 266
3. 1.6	Joe is 8 points above the mean. $\frac{8}{5} = 1.6$
4. 70	$50 + (2 \times 10)$
5. 16 percent	

Reliability and Validity

277. The statistical analogies in the discussion below are in parentheses.

A rifle placed in a vise might hit the same place consistently:

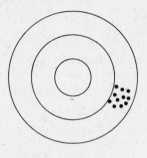

(This is reliability.)

In addition, the rifle might be on target:

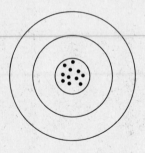

(This is validity.)

Could the rifle be on target (valid) if it wasn't consistent (reliable)? _____

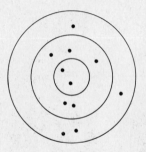

No

278. Reliability is expressed as a correlation coefficient that measures the self-consistency of a test. If any structured method (instrument, test, experiment, etc.) is reliable, the same results should be obtainable time after time. If one were to use a very crude scale for weighing chemicals, he could not expect his results to be _____.

Reliable

279. The three most common methods of measuring reliability are test-retest, split-half, and equivalent forms. If a test is to give consistent results, it has to be _____.

Reliable

280. Test-retest reliability is established by correlating the scores on the *same test given at two different times.*
In the illustration below, would the form on the right be administered on May 1? _____

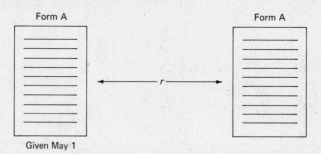

Form A   *r*   Form A

Given May 1

No

281. Equivalent forms reliability is established by the correlation between *equivalent forms of the same test given at the same time.*
In the illustration below, would the form on the right be administered on May 1? _____

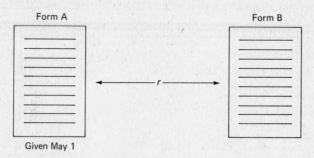

Form A   *r*   Form B

Given May 1

Yes

Note: Researchers often measure equivalence and stability (test-retest) in combination, by giving equivalent forms at two different times.

282. Split-half reliability is established by correlating the scores on *two halves of the same test given at the same time.*
This type of reliability is most logically related to which of the other two types? _____

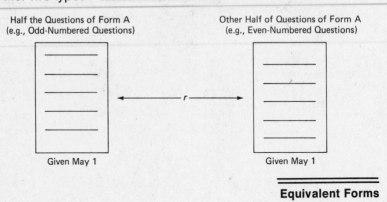

Half the Questions of Form A
(e.g., Odd-Numbered Questions)

Other Half of Questions of Form A
(e.g., Even-Numbered Questions)

Given May 1

Given May 1

Equivalent Forms

283. Validity is expressed as the *extent to which a test measures what it is supposed to measure* (i.e., how well it hits the target). One does not ask if a test is *valid,* but for *what* is it _____?

Valid

284. There are four types of validity with which the student should be concerned. They are *predictive, concurrent, content,* and *construct.* Each of these types of validity indicates the test's usefulness for some particular purpose. If we want to know how useful a test is, we have to know how _____ it is for some particular purpose.

Valid

285. Validity is the correlation between a test result and some performance called a *criterion.* (This will not be true for content validity.)
The performance against which a test is matched is called the _____.

Criterion

286. A measure of how well a test predicts a person's future job or school performance is called *predictive validity.* In predictive validity the performance predicted by the test is the _____ performance.

Criterion (or Future)

287. If a test were designed to predict grade averages for incoming freshmen, the criterion would be the actual grade average obtained at the end of the freshman year. The test's efficiency at this task would be called its _____ _____.

Predictive Validity

288. One might be interested in how well a test measured some *current* situation. If high scorers on a leadership test are indeed leaders—as simultaneously determined by judges' ratings—that test could be said to have *concurrent* _____.

Validity

289. The only difference between concurrent and predictive validity is time. A test with concurrent validity seeks to estimate *present* performance whereas a test with predictive validity seeks to estimate _____ performance.

Future

290. It is easy to understand a third type of validity, construct validity, when one understands the psychological meaning of the word *construct.* A construct ties together a series of related observations. For example, a person might laugh a lot, whistle frequently, speak in resonant tones, and move with unusual vitality. These are all physical observations but if one puts them together and says a person is "happy," one has formed a *construct.* Similarly, schizophrenia, generosity, goodness, objectivity, maturity, etc. are not simple acts but are _____ derived from a whole set of simple acts.

Constructs

291. The correlation between a test for determining the degree of depression and the clinical observations of simple and separate acts indicating depression would reflect the test's _____ and _____ validity.

Construct   Concurrent

292. The synthesis derived from the clinical observations serves as the _____.

Criterion

293. If a test that had been designed to predict which students would develop *schizophrenia* by their senior year (provided treatment was not sought) correleated very well with *subsequent clinical* findings, one could say that the test had both _____ and _____ validity.

Predictive   Construct

294. An examination for a course should reflect the content of the course. The degree to which an examination tests the *content* of a course is called _____ validity.

Content

295. If the lectures in a course were mostly on French political history but the greatest percentage of exam questions inquired about English economic theory, one could say the examination had low _____ _____.

Content Validity

296. In order for a test to have a high predictive, concurrent, or construct validity, it must first of all be *reliable.* A test could not be used to predict future performance if the test were not _____.

Reliable

297. From the above it is apparent that reliability can be expressed by a coefficient of correlation representing the relationship between two tests. Validity, however, is expressed by a correlation coefficient representing the relationship between a test and some criterion. The correlation coefficient between a set of measurements and a criterion indicates the degree of _____.

═══
Validity

Standard Error of Measurement

298. Previous experience with this statistics program has no doubt led the reader to the conclusion that an individual test score is something about which one can logically be skeptical. For example, the law of mean regression discussed previously suggests that high or low instances in a distribution, upon retesting, tend to move toward the _____ of the distribution.

═══
Mean

299. Because every measurement has some unreliability, one can never be certain that one particular administration of a test, for instance, yields the "true" value or score for that individual. If it's not possible to know the "true" value or score, it's at least desirable to know how far off we might be. This is possible by the use of the standard error of measurement. Does the standard error of measurement allow us to determine the "true" value or score? _____

══
No

300. The concept of standard error of measurement assumes that the "true" value or score for an individual would be the mean of an infinite number of times he took the same test, with the assumption that each time he took the test it was as if he had never taken the test before. Let us be very clear that we are

talking all the time here about _____ person taking
_____ test _____ _____
_____ of times. We could list each of one individual's scores and draw a curve to fit this population of his
test scores. Let us say he is taking the same IQ test a million
times. His various scores on this test might yield a curve such
as that shown below.

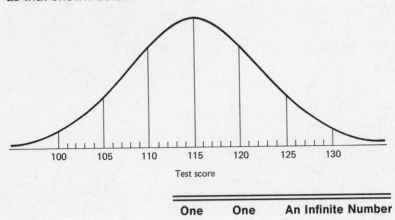

Test score

301. The same percentages apply here as in the standard deviation
     discussion. Thus, 68 percent of this one person's scores on this
     one test, taken an infinite number of times, would fall within one
     SD of his mean score for all the times he took this one test.
     What percentage of his scores would fall within 2 SD of the
     mean (which we may call his "true" score on this particular
     test)? _____

302. The mean, 115, is called his "true" score. Could he ever get 130
     on this same test? _____ Could he ever get
     100 on this same test? _____

303. If a person had a Binet IQ of 120, and the standard error of the
     test was 5 IQ units, the student might be told that there was a
     95 percent chance (you remember this as the area between

114

−2σ and +2σ) that his "true" score was between 120 − (2 × 5) = 110 and _____.

====
130

304. This particular administration might be one of the dots at the low end of a distribution centered about a "true" value or score of 130 (see A below), or it might be at the high end of a distribution centered about a "true" value or score of 110 (see B below).

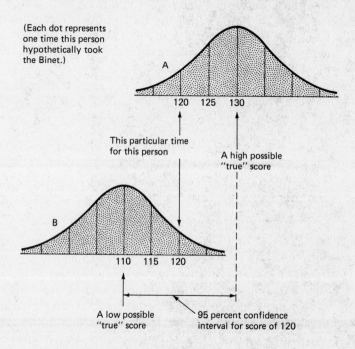

(Each dot represents one time this person hypothetically took the Binet.)

A

120 125 130

This particular time for this person

A high possible "true" score

B

110 115 120

A low possible "true" score

95 percent confidence interval for score of 120

By using the standard error of the mean, do we know the direction in which the true mean lies? _____

====
No

305. If one knows the standard deviation and the reliability coefficient of a test, it is easy to determine its *standard error* of measurement by the following formula: *standard error of measurement = standard deviation* × $\sqrt{1-r}$. In the preceding equation, *r* is the reliability coefficient. What is the standard

error of measurement for a test with an SD of 12 and an internal reliability of .75? _____

$$= 6$$

306. The band within which one can feel a certain confidence that a score or instance lies is called a confidence interval. The 95 percent confidence interval of the person with the Binet IQ of 120 was thus _____ to _____.
Assuming that this standard error was correct, would it be safe (and by safe here is meant only being wrong a maximum of 1 time out of every 20 or 5 percent of the time) to say that a person with a Binet IQ of 130 was brighter than a person with a Binet IQ of 119? _____

| 110 | 130 | No |

307. When one is presented with some individual measurement, such as an SAT score of 450 or an examination grade of 69, one's mind's eye should not visualize a specific point on a continum as shown below.

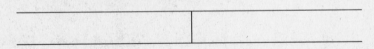

But one should visualize, rather, a hazy band. This hazy band typically extends in either direction from the same position. However, the true score may be located in a fairly dense area or in a sparse area. Typically, the true score is in the more dense areas and one can visualize the figure below with the single reported score occurring in the center. The dots representing possible true scores thin out in areas of limited occurrence at the ±3 standard error (SE) positions located on appropriate sides of the reported score.

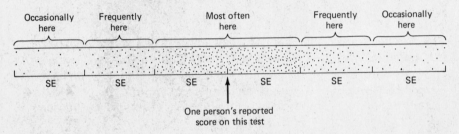

116

Hence the standard error of measurement tells one the
_____ limits surrounding a given test score.
These limits define the space within which a person's "true"
score has a given _____ of occurring.

Confidence (or Probable)	Probability

OPEN BOOK QUIZ

1. The correlation between an aptitude test and the achievement it
was being used to predict is a measure of _____.

2. If a test gave very reproducible results, it would be described as

_____.

3. If a test had a standard deviation of 10 and a reliability of .64,
what would be its standard error of measurement? _____.

4. If a person got a score of 550 on a test with a standard error of
measurement equal to 15, we could be 95 percent confident that
his true score lay between _____ and _____.

5. Of the three common methods of measuring reliability, which
expresses the correlation between the score on the same set of
items given at two different times? _____.

Answers	Comments
1. Validity (or predictive validity)	
2. Reliable	
3. 6	$SD \times \sqrt{1-r} = 10\sqrt{1-.64}$
4. 520 and 580	$550 + (2 \times 15)$
5. Test-retest	See Frame 280

Diagnostic Test and Alphabetical Index

Instructions

This test will help you diagnose any weak areas in your understanding of the main statistical concepts discussed in the program. Also, it will serve as an exercise in recall and review.

Work these exercises, checking your answers as you go. (Answers are given on the back of each page.) If you miss any questions and can't figure out the reason for the answers, you will find the appropriate beginning frame number opposite the key index word in the right-hand column on the page. Turn to the beginning frame and rework the appropriate section until you are certain that you have mastered the concept.

If you are then still confused, seek assistance from your course instructor.

Diagnostic Test	Alphabetical Index	Initial Frame
1. Which axis is the abscissa? Y X _____	**Abscissa**	7
2. An analysis of variance could be used to test the significance of differences among _____.	**Analysis of variance**	158
3. Give three measures of central tendency: _____, _____, _____.	**Central tendency**	10
4. A chi-square test determines the significance of deviations from expected _____.	**Chi-square test**	211
5. A confidence interval is a range within which a population _____ would most likely fall.	**Confidence interval**	146 231 306
6. Correlation coefficients may range all the way from _____ to _____.	**Correlation coefficient**	110

Answers

1. X
2. Means
3. Mean, Median, Mode
4. Frequencies
5. Parameter
6. -1.0 to $+1.0$

Diagnostic Test	Alphabetical Index	Initial Frame
7. *f* is an abbreviation for _____.	**f**	8
8. A frequency distribution is a _____ of how many times each value occurs.	**Frequency distribution**	1
9. For the distribution 1, 2, 2, 3, 3, 3, 3, 4, 5, draw a frequency polygon.	**Frequency polygon**	5
10. For the distribution 1, 2, 2, 3, 3, 3, 3, 4, 5, draw a histogram.	**Histogram**	6
11. Inferential statistics are used to _____ _____ _____ unknown data from known data.	**Inferential statistics**	131
12. For the distribution 0, 2, 2, 3, 5, 6, give the following: a. Mean ____ b. Median ____ c. Mode ____	**Mean** **Median** **Mode**	11 33 42

Answers

7. Frequency
8. Tally or Count
9.

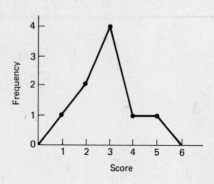

10.

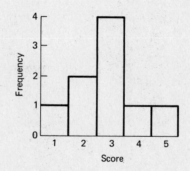

11. Infer estimates of (or Prophesy)
12.
   a. Mean = 3
   b. Median = 2.5
   c. Mode = 2

13. What percentage (to the nearest whole percent) of the normal curve area lies within A? _____ Within Within B? _____ Within C? _____

Normal curve— areas under the curve

70

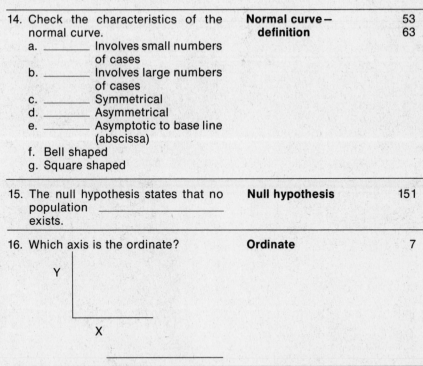

Standard deviations

14. Check the characteristics of the normal curve.
    a. _____ Involves small numbers of cases
    b. _____ Involves large numbers of cases
    c. _____ Symmetrical
    d. _____ Asymmetrical
    e. _____ Asymptotic to base line (abscissa)
    f. Bell shaped
    g. Square shaped

Normal curve— definition

53
63

15. The null hypothesis states that no population _____ exists.

Null hypothesis

151

16. Which axis is the ordinate?

Ordinate

7

Y

X

_____

Answers

13.
    A: 68
    B: 95
    C: 100
14.
    b. Involves Large Numbers of Cases
    c. Symmetrical
    e. Asymptotic to Base Line (Abscissa)
    f. Bell Shaped
15. Difference
16. Y

Diagnostic Test	Alphabetical Index	Initial Frame
17. Parameter is used to refer to a _____ value.	**Parameter**	141
18. In the distribution 0, 2, 3, 4, 6, what is the percentile rank of the score 3? _____	**Percentile rank**	239
19. A population is all of something about which an _____ is to be made.	**Population**	131
20. For the distribution 0, 2, 3, 4, 6, give the range. _____ Is it more (or less) stable than the standard deviation? _____	**Range (compared with standard deviation)**	84 106
21. If the correlation between a predictor variable and a future performance is low, the best guess for any particular person's future performance will be (quite near or quite far from) _____ _____ the mean future performance for all persons.	**Regression**	187
22. The most common index of relationship between two variables is the _____.	**Relationship**	110
23. After each of the (three) reliability types below, place two letters indicating which of the following characteristics apply: (a) Given at same time (b) Given at different time (c) Uses same test over (d) Uses similar form of test (e) Uses an internal comparison 1. Test-retest ____ ____ 2. Equivalent forms ____ ____ 3. Split-half ____ ____	**Reliability**	277

Answers

17. Population
18. 50
19. Estimate
20. 6; Less
21. Quite Near
22. Correlation Coefficient
23.
    1. b, c
    2. a, d (sometimes b)
    3. a, e

Diagnostic Test	Alphabetical Index	Initial Frame
24. A sample is a portion of a _____.	**Sample**	131
25. When one is using the null hypothesis, if a sample difference is believed to reflect a true population _____, the sample difference is said to have significance.	**Significance**	147
26. For the distribution 0, 2, 2, 2, 5, 40, the skewness is _____ (positive or negative).	**Skewness**	16
27. For the distribution 0, 2, 3, 4, 6, give the standard deviation. _____ The standard deviation is the square root of the _____.	**Standard deviation**	91 (For background, See Frames 64–90)
28. If a person received a score of 120 on a test with a standard error of 5, one could feel 95 percent confident that his true score lay between _____ and _____.	**Standard error of measurement**	298
29. Facile interpretation of standard scores cannot occur unless the distribution of the population is _____.	**Standard score**	264
30. Statistic may be used to refer to a _____ value.	**Statistic**	141
31. For the distribution 0, 2, 3, 4, 6, compute the sum of squares.	**Sum of squares**	101

Answers

24. Population
25. Difference
26. Positive
27. 2; Variance
28. 110 and 130
29. Normal
30. Sample
31. 20

Diagnostic Test	Alphabetical Index	Initial Frame
32. Use the numbers below to indicate which of the following types of scores a. Are relative to a group's performance _____ b. Is easiest to obtain _____ c. Is most meaningful when the distribution is normal _____ (1) Raw score (2) Percentile rank (3) Standard score	**Test scores— interpretation of**	241
33. Fill in the types of validity expressing a test's ability to do the following: a. Foretell a future event _____ b. Reflect a present situation _____ c. Reflect that which was actually covered in a course _____ d. Accurately label a set of related simple acts _____	**Validity**	277
34. Give two measures of variability. a. _____; _____ Which is the most reliable? b. _____	**Variability**	83
35. If the raw scores in a population had a mean of 60 and a standard deviation of 2, a raw score of 58 would have a z score of _____.	**z Score**	267

Answers

32.
    a. 2, 3
    b. 1
    c. 3
33.
    a. Predictive
    b. Concurrent
    c. Content
    d. Construct
34.
    a. Range;
       Standard Deviation or Variance
    b. Standard Deviation or Variance
35. −1

Appendix

Tables of Critical Values

Table A   Critical values of the correlation coefficient

Note: For directions on the use of this table, see pages 57–58:

df	Level of Significance .10	.05	.02	.01
1	.9877	.9969	.9995	.9999
2	.9000	.9500	.9800	.9900
3	.8054	.8783	.9343	.9587
4	.7293	.8114	.8822	.9172
5	.6694	.7545	.8329	.8745
6	.6215	.7067	.7887	.8343
7	.5822	.6664	.7498	.7977
8	.5494	.6319	.7155	.7646
9	.5214	.6021	.6851	.7348
10	.4973	.5760	.6581	.7079
11	.4762	.5529	.6339	.6835
12	.4575	.5324	.6120	.6614
13	.4409	.5139	.5923	.6411
14	.4259	.4973	.5742	.6226
15	.4124	.4821	.5577	.6055
16	.4000	.4683	.5425	.5897
17	.3887	.4555	.5285	.5751
18	.3783	.4438	.5155	.5614
19	.3687	.4329	.5034	.5487
20	.3598	.4227	.4921	.5368
25	.3233	.3809	.4451	.4869
30	.2960	.3494	.4093	.4487
35	.2746	.3246	.3810	.4182
40	.2573	.3044	.3578	.3932
45	.2428	.2875	.3384	.3721
50	.2306	.2732	.3218	.3541
60	.2108	.2500	.2948	.3248
70	.1954	.2319	.2737	.3017
80	.1829	.2172	.2565	.2830
90	.1726	.2050	.2422	.2673
100	.1638	.1946	.2301	.2540

Abridged from Table VI of Fisher and Yates: "Statistical Tables for Biological, Agricultural and Medical Research," published by Oliver and Boyd Ltd., Edinburgh, and by permission of the authors and publishers.

Table B The 5 percent points for the distribution of F

Note: For directions on the use of this table, see pages 67–68.

Denominator df	1	2	3	4	5	6	8	12	24	∞
1	161.4	199.5	215.7	224.6	230.2	234.0	238.9	243.9	249.0	254.3
2	18.51	19.00	19.16	19.25	19.30	19.33	19.37	19.41	19.45	19.50
3	10.13	9.55	9.28	9.12	9.01	8.94	8.84	8.74	8.64	8.53
4	7.71	6.94	6.59	6.39	6.26	6.16	6.04	5.91	5.77	5.63
5	6.61	5.79	5.41	5.19	5.05	4.95	4.82	4.68	4.53	4.36
6	5.99	5.14	4.76	4.53	4.39	4.28	4.15	4.00	3.84	3.67
7	5.59	4.74	4.35	4.12	3.97	3.87	3.73	3.57	3.41	3.23
8	5.32	4.46	4.07	3.84	3.69	3.58	3.44	3.28	3.12	2.93
9	5.12	4.26	3.86	3.63	3.48	3.37	3.23	3.07	2.90	2.71
10	4.96	4.10	3.71	3.48	3.33	3.22	3.07	2.91	2.74	2.54
11	4.84	3.98	3.59	3.36	3.20	3.09	2.95	2.79	2.61	2.40
12	4.75	3.88	3.49	3.26	3.11	3.00	2.85	2.69	2.50	2.30
13	4.67	3.80	3.41	3.18	3.02	2.92	2.77	2.60	2.42	2.21
14	4.60	3.74	3.34	3.11	2.96	2.85	2.70	2.53	2.35	2.13
15	4.54	3.68	3.29	3.06	2.90	2.79	2.64	2.48	2.29	2.07
16	4.49	3.63	3.24	3.01	2.85	2.74	2.59	2.42	2.24	2.01
17	4.45	3.59	3.20	2.96	2.81	2.70	2.55	2.38	2.19	1.96
18	4.41	3.55	3.16	2.93	2.77	2.66	2.51	2.34	2.15	1.92
19	4.38	3.52	3.13	2.90	2.74	2.63	2.48	2.31	2.11	1.88
20	4.35	3.49	3.10	2.87	2.71	2.60	2.45	2.28	2.08	1.84
21	4.32	3.47	3.07	2.84	2.68	2.57	2.42	2.25	2.05	1.81
22	4.30	3.44	3.05	2.82	2.66	2.55	2.40	2.23	2.03	1.78
23	4.28	3.42	3.03	2.80	2.64	2.53	2.38	2.20	2.00	1.76
24	4.26	3.40	3.01	2.78	2.62	2.51	2.36	2.18	1.98	1.73
25	4.24	3.38	2.99	2.76	2.60	2.49	2.34	2.16	1.96	1.71
26	4.22	3.37	2.98	2.74	2.59	2.47	2.32	2.15	1.95	1.69
27	4.21	3.35	2.96	2.73	2.57	2.46	2.30	2.13	1.93	1.67
28	4.20	3.34	2.95	2.71	2.56	2.44	2.29	2.12	1.91	1.65
29	4.18	3.33	2.93	2.70	2.54	2.43	2.28	2.10	1.90	1.64
30	4.17	3.32	2.92	2.69	2.53	2.42	2.27	2.09	1.89	1.62
40	4.08	3.23	2.84	2.61	2.45	2.34	2.18	2.00	1.79	1.51
60	4.00	3.15	2.76	2.52	2.37	2.25	2.10	1.92	1.70	1.39
120	3.92	3.07	2.68	2.45	2.29	2.17	2.02	1.83	1.61	1.25
∞	3.84	2.99	2.60	2.37	2.21	2.10	1.94	1.75	1.52	1.00

Numerator df

Note: In using this table, the greater mean square must be the numerator of F.

Abridged from Table V of Fisher and Yates: "Statistical Tables for Biological, Agricultural and Medical Research," published by Oliver and Boyd Ltd., Edinburgh, and by permission of the authors and publishers.

Table C   The 1 percent points for the distribution of F

Note: For directions on the use of this table, see pages 67–68.

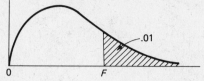

Denominator df	Numerator df									
	1	*2*	*3*	*4*	*5*	*6*	*8*	*12*	*24*	∞
1	4052	4999	5403	5625	5764	5859	5982	6106	6234	6366
2	98.50	99.00	99.17	99.25	99.30	99.33	99.37	99.42	99.46	99.50
3	34.12	30.82	29.46	28.71	28.24	27.91	27.49	27.05	26.60	26.12
4	21.20	18.00	16.69	15.98	15.52	15.21	14.80	14.37	13.93	13.46
5	16.26	13.27	12.06	11.39	10.97	10.67	10.29	9.89	9.47	9.02
6	13.74	10.92	9.78	9.15	8.75	8.47	8.10	7.72	7.31	6.88
7	12.25	9.55	8.45	7.85	7.46	7.19	6.84	6.47	6.07	5.65
8	11.26	8.65	7.59	7.01	6.63	6.37	6.03	5.67	5.28	4.86
9	10.56	8.02	6.99	6.42	6.06	5.80	5.47	5.11	4.73	4.31
10	10.04	7.56	6.55	5.99	5.64	5.39	5.06	4.71	4.33	3.91
11	9.65	7.20	6.22	5.67	5.32	5.07	4.74	4.40	4.02	3.60
12	9.33	6.93	5.95	5.41	5.06	4.82	4.50	4.16	3.78	3.36
13	9.07	6.70	5.74	5.20	4.86	4.62	4.30	3.96	3.59	3.16
14	8.86	6.51	5.56	5.03	4.69	4.46	4.14	3.80	3.43	3.00
15	8.68	6.36	5.42	4.89	4.56	4.32	4.00	3.67	3.29	2.87
16	8.53	6.23	5.29	4.77	4.44	4.20	3.89	3.55	3.18	2.75
17	8.40	6.11	5.18	4.67	4.34	4.10	3.79	3.45	3.08	2.65
18	8.28	6.01	5.09	4.58	4.25	4.01	3.71	3.37	3.00	2.57
19	8.18	5.93	5.01	4.50	4.17	3.94	3.63	3.30	2.92	2.49
20	8.10	5.85	4.94	4.43	4.10	3.87	3.56	3.23	2.86	2.42
21	8.02	5.78	4.87	4.37	4.04	3.81	3.51	3.17	2.80	2.36
22	7.94	5.72	4.82	4.31	3.99	3.76	3.45	3.12	2.75	2.31
23	7.88	5.66	4.76	4.26	3.94	3.71	3.41	3.07	2.70	2.26
24	7.82	5.61	4.72	4.22	3.90	3.67	3.36	3.03	2.66	2.21
25	7.77	5.57	4.68	4.18	3.86	3.63	3.32	2.99	2.62	2.17
26	7.72	5.53	4.64	4.14	3.82	3.59	3.29	2.96	2.58	2.13
27	7.68	5.49	4.60	4.11	3.78	3.56	3.26	2.93	2.55	2.10
28	7.64	5.45	4.57	4.07	3.75	3.53	3.23	2.90	2.52	2.06
29	7.60	5.42	4.54	4.04	3.73	3.50	3.20	2.87	2.49	2.03
30	7.56	5.39	4.51	4.02	3.70	3.47	3.17	2.84	2.47	2.01
40	7.31	5.18	4.31	3.83	3.51	3.29	2.99	2.66	2.29	1.80
60	7.08	4.98	4.13	3.65	3.34	3.12	2.82	2.50	2.12	1.60
120	6.85	4.79	3.95	3.48	3.17	2.96	2.66	2.34	1.95	1.38
∞	6.64	4.60	3.78	3.32	3.02	2.80	2.51	2.18	1.79	1.00

Note: In using this table, the greater mean square must be the numerator of F.

Abridged from Table V of Fisher and Yates: "Statistical Tables for Biological, Agricultural and Medical Research," published by Oliver and Boyd Ltd., Edinburgh, and by permission of the authors and publishers.

Table D Distribution of *t* probability

Note: For directions on the use of this table, see page 95.

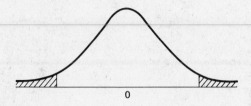

df	.1	.05	.01	.001
1	6.314	12.706	63.657	636.619
2	2.920	4.303	9.925	31.598
3	2.353	3.182	5.841	12.941
4	2.132	2.776	4.604	8.610
5	2.015	2.571	4.032	6.859
6	1.943	2.447	3.707	5.959
7	1.895	2.365	3.499	5.405
8	1.860	2.306	3.355	5.041
9	1.833	2.262	3.250	4.781
10	1.812	2.228	3.169	4.587
11	1.796	2.201	3.106	4.437
12	1.782	2.179	3.055	4.318
13	1.771	2.160	3.012	4.221
14	1.761	2.145	2.977	4.140
15	1.753	2.131	2.947	4.073
16	1.746	2.120	2.921	4.015
17	1.740	2.110	2.898	3.965
18	1.734	2.101	2.878	3.922
19	1.729	2.093	2.861	3.883
20	1.725	2.086	2.845	3.850
21	1.721	2.080	2.831	3.819
22	1.717	2.074	2.819	3.792
23	1.714	2.069	2.807	3.767
24	1.711	2.064	2.797	3.745
25	1.708	2.060	2.787	3.725
26	1.706	2.056	2.779	3.707
27	1.703	2.052	2.771	3.690
28	1.701	2.048	2.763	3.674
29	1.699	2.045	2.756	3.659
30	1.697	2.042	2.750	3.646
40	1.684	2.021	2.704	3.551
60	1.671	2.000	2.660	3.460
120	1.658	1.980	2.617	3.373
∞	1.645	1.960	2.576	3.291

Abridged from Table III of R. A. Fisher and F. Yates: "Statistical Tables for Biological, Agricultural, and Medical Research," published by Oliver and Boyd Ltd., Edinburgh. Abridged with permission of the authors and publisher.

Table E Distribution of χ^2

Note: For directions on the use of this table, see page 82.

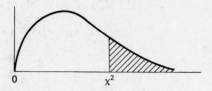

df	P				
	.10	.05	.02	.01	.001
1	2.706	3.841	5.412	6.635	10.827
2	4.605	5.991	7.824	9.210	13.815
3	6.251	7.815	9.837	11.345	16.266
4	7.779	9.488	11.668	13.277	18.467
5	9.236	11.070	13.388	15.086	20.515
6	10.645	12.592	15.033	16.812	22.457
7	12.017	14.067	16.622	18.475	24.322
8	13.362	15.507	18.168	20.090	26.125
9	14.684	16.919	19.679	21.666	27.877
10	15.987	18.307	21.161	23.209	29.588
11	17.275	19.675	22.618	24.725	31.264
12	18.549	21.026	24.054	26.217	32.909
13	19.812	22.362	25.472	27.688	34.528
14	21.064	23.685	26.873	29.141	36.123
15	22.307	34.996	28.259	30.578	37.697
16	23.542	26.296	29.633	32.000	39.252
17	24.769	27.587	30.995	33.409	40.790
18	25.989	28.869	32.346	34.805	42.312
19	27.204	30.144	33.687	36.191	43.820
20	28.412	31.410	35.020	37.566	45.315
21	29.615	32.671	36.343	38.932	46.797
22	30.813	33.924	37.659	40.289	48.268
23	32.007	35.172	38.968	41.638	49.728
24	33.196	36.415	40.270	42.980	51.179
25	34.382	37.652	41.566	44.314	52.620
26	35.563	38.885	42.856	45.642	54.052
27	36.741	40.113	44.140	46.963	55.476
28	37.916	41.337	45.419	48.278	56.893
29	39.087	42.557	46.693	49.588	58.302
30	40.256	43.773	47.692	50.892	59.703

Abridged from Table IV of Fisher and Yates: "Statistical Tables for Biological, Agricultural and Medical Research," published by Oliver and Boyd Ltd., Edinburgh, and by permission of the authors and publishers.

Table F Areas and Ordinates of the Normal Curve in Terms of z

Note: For directions on the use of this table, see pages 24–32.

(1) z Standard Score $\left(\frac{x}{\sigma}\right)$	(2) A Area from Mean to z	(3) B Area in Larger Portion	(4) C Area in Smaller Portion	(5) y Ordinate at z
0.00	.0000	.5000	.5000	.3989
0.01	.0040	.5040	.4960	.3989
0.02	.0080	.5080	.4920	.3989
0.03	.0120	.5120	.4880	.3988
0.04	.0160	.5160	.4840	.3986
0.05	.0199	.5199	.4801	.3984
0.06	.0239	.5239	.4761	.3982
0.07	.0279	.5279	.4721	.3980
0.08	.0319	.5319	.4681	.3977
0.09	.0359	.5359	.4641	.3973
0.10	.0398	.5398	.4602	.3970
0.11	.0438	.5438	.4562	.3965
0.12	.0478	.5478	.4522	.3961
0.13	.0517	.5517	.4483	.3956
0.14	.0557	.5557	.4443	.3951
0.15	.0596	.5596	.4404	.3945
0.16	.0636	.5636	.4364	.3939
0.17	.0675	.5675	.4325	.3932
0.18	.0714	.5714	.4286	.3925
0.19	.0753	.5753	.4247	.3918
0.20	.0793	.5793	.4207	.3910
0.21	.0832	.5832	.4168	.3902
0.22	.0871	.5871	.4129	.3894
0.23	.0910	.5910	.4090	.3885
0.24	.0948	.5948	.4052	.3876
0.25	.0987	.5987	.4013	.3867
0.26	.1026	.6026	.3974	.3857
0.27	.1064	.6064	.3936	.3847
0.28	.1103	.6103	.3897	.3836
0.29	.1141	.6141	.3859	.3825
0.30	.1179	.6179	.3821	.3814
0.31	.1217	.6217	.3783	.3802
0.32	.1255	.6255	.3745	.3790
0.33	.1293	.6293	.3707	.3778
0.34	.1331	.6331	.3669	.3765
0.35	.1368	.6368	.3632	.3752
0.36	.1406	.6406	.3594	.3739

From *Statistical Methods for the Behavioral Sciences* by Allen L. Edwards. Copyrighted 1954, by Allen L. Edwards. Reprinted by permission of Holt, Rinehart and Winston, Publishers.

0.37	.1443	.6443	.3557	.3725
0.38	.1480	.6480	.3520	.3712
0.39	.1517	.6517	.3483	.3697
0.40	.1554	.6554	.3446	.3683
0.41	.1591	.6591	.3409	.3668
0.42	.1628	.6628	.3372	.3653
0.43	.1664	.6664	.3336	.3637
0.44	.1700	.6700	.3300	.3621
0.45	.1736	.6736	.3264	.3605
0.46	.1772	.6772	.3228	.3589
0.47	.1808	.6808	.3192	.3572
0.48	.1844	.6844	.3156	.3555
0.49	.1879	.6879	.3121	.3538
0.50	.1915	.6915	.3085	.3521
0.51	.1950	.6950	.3050	.3503
0.52	.1985	.6985	.3015	.3485
0.53	.2019	.7019	.2981	.3467
0.54	.2054	.7054	.2946	.3448
0.55	.2088	.7088	.2912	.3429
0.56	.2123	.7123	.2877	.3410
0.57	.2157	.7157	.2843	.3391
0.58	.2190	.7190	.2810	.3372
0.59	.2224	.7224	.2776	.3352
0.60	.2257	.7257	.2743	.3332
0.61	.2291	.7291	.2709	.3312
0.62	.2324	.7324	.2676	.3292
0.63	.2357	.7357	.2643	.3271
0.64	.2389	.7389	.2611	.3251
0.65	.2422	.7422	.2578	.3230
0.66	.2454	.7454	.2546	.3209
0.67	.2486	.7486	.2514	.3187
0.68	.2517	.7517	.2483	.3166
0.69	.2549	.7549	.2451	.3144
0.70	.2580	.7580	.2420	.3123
0.71	.2611	.7611	.2389	.3101
0.72	.2642	.7642	.2358	.3079
0.73	.2673	.7673	.2327	.3056
0.74	.2704	.7704	.2296	.3034
0.75	.2734	.7734	.2266	.3011
0.76	.2764	.7764	.2236	.2989
0.77	.2794	.7794	.2206	.2966
0.78	.2823	.7823	.2177	.2943
0.79	.2852	.7852	.2148	.2920
0.80	.2881	.7881	.2119	.2897
0.81	.2910	.7910	.2090	.2874
0.82	.2939	.7939	.2061	.2850
0.83	.2967	.7967	.2033	.2827
0.84	.2995	.7995	.2005	.2803
0.85	.3023	.8023	.1977	.2780
0.86	.3051	.8051	.1949	.2756
0.87	.3078	.8078	.1922	.2732
0.88	.3106	.8106	.1894	.2709
0.89	.3133	.8133	.1867	.2685

Table F (cont.)

(1) z Standard Score $\left(\frac{x}{\sigma}\right)$	(2) A Area from Mean to z	(3) B Area in Larger Portion	(4) C Area in Smaller Portion	(5) y Ordinate at z
0.90	.3159	.8159	.1841	.2661
0.91	.3186	.8186	.1814	.2637
0.92	.3212	.8212	.1788	.2613
0.93	.3238	.8238	.1762	.2589
0.94	.3264	.8264	.1736	.2565
0.95	.3289	.8289	.1711	.2541
0.96	.3315	.8315	.1685	.2516
0.97	.3340	.8340	.1660	.2492
0.98	.3365	.8365	.1635	.2468
0.99	.3389	.8389	.1611	.2444
1.00	.3413	.8413	.1587	.2420
1.01	.3438	.8438	.1562	.2396
1.02	.3461	.8461	.1539	.2371
1.03	.3485	.8485	.1515	.2347
1.04	.3508	.8508	.1492	.2323
1.05	.3531	.8531	.1469	.2299
1.06	.3554	.8554	.1446	.2275
1.07	.3577	.8577	.1423	.2251
1.08	.3599	.8599	.1401	.2227
1.09	.3621	.8621	.1379	.2203
1.10	.3643	.8643	.1357	.2179
1.11	.3665	.8665	.1335	.2155
1.12	.3686	.8686	.1314	.2131
1.13	.3708	.8708	.1292	.2107
1.14	.3729	.8729	.1271	.2083
1.15	.3749	.8749	.1251	.2059
1.16	.3770	.8770	.1230	.2036
1.17	.3790	.8790	.1210	.2012
1.18	.3810	.8810	.1190	.1989
1.19	.3830	.8830	.1170	.1965
1.20	.3849	.8849	.1151	.1942
1.21	.3869	.8869	.1131	.1919
1.22	.3888	.8888	.1112	.1895
1.23	.3907	.8907	.1093	.1872
1.24	.3925	.8925	.1075	.1849
1.25	.3944	.8944	.1056	.1826
1.26	.3962	.8962	.1038	.1804
1.27	.3980	.8980	.1020	.1781
1.28	.3997	.8997	.1003	.1758
1.29	.4015	.9015	.0985	.1736
1.30	.4032	.9032	.0968	.1714
1.31	.4049	.9049	.0951	.1691
1.32	.4066	.9066	.0934	.1669

142

1.33	.4082	.9082	.0918	.1647
1.34	.4099	.9099	.0901	.1626
1.35	.4115	.9115	.0885	.1604
1.36	.4131	.9131	.0869	.1582
1.37	.4147	.9147	.0853	.1561
1.38	.4162	.9162	.0838	.1539
1.39	.4177	.9177	.0823	.1518
1.40	.4192	.9192	.0808	.1497
1.41	.4207	.9207	.0793	.1476
1.42	.4222	.9222	.0778	.1456
1.43	.4236	.9236	.0764	.1435
1.44	.4251	.9251	.0749	.1415
1.45	.4265	.9265	.0735	.1394
1.46	.4279	.9279	.0721	.1374
1.47	.4292	.9292	.0708	.1354
1.48	.4306	.9306	.0694	.1334
1.49	.4319	.9319	.0681	.1315
1.50	.4332	.9332	.0668	.1295
1.51	.4345	.9345	.0655	.1276
1.52	.4357	.9357	.0643	.1257
1.53	.4370	.9370	.0630	.1238
1.54	.4382	.9382	.0618	.1219
1.55	.4394	.9394	.0606	.1200
1.56	.4406	.9406	.0594	.1182
1.57	.4418	.9418	.0582	.1163
1.58	.4429	.9429	.0571	.1145
1.59	.4441	.9441	.0559	.1127
1.60	.4452	.9452	.0548	.1109
1.61	.4463	.9463	.0537	.1092
1.62	.4474	.9474	.0526	.1074
1.63	.4484	.9484	.0516	.1057
1.64	.4495	.9495	.0505	.1040
1.65	.4505	.9505	.0495	.1023
1.66	.4515	.9515	.0485	.1006
1.67	.4525	.9525	.0475	.0989
1.68	.4535	.9535	.0465	.0973
1.69	.4545	.9545	.0455	.0957
1.70	.4554	.9554	.0446	.0940
1.71	.4564	.9564	.0436	.0925
1.72	.4573	.9573	.0427	.0909
1.73	.4582	.9582	.0418	.0893
1.74	.4591	.9591	.0409	.0878
1.75	.4599	.9599	.0401	.0863
1.76	.4608	.9608	.0392	.0848
1.77	.4616	.9616	.0384	.0833
1.78	.4625	.9625	.0375	.0818
1.79	.4633	.9633	.0367	.0804
1.80	.4641	.9641	.0359	.0790
1.81	.4649	.9649	.0351	.0775
1.82	.4656	.9656	.0344	.0761
1.83	.4664	.9664	.0336	.0748
1.84	.4671	.9671	.0329	.0734
1.85	.4648	.9678	.0322	.0721
1.86	.4686	.9686	.0314	.0707

(1) z Standard Score $\left(\frac{x}{\sigma}\right)$	(2) A Area from Mean to z	(3) B Area in Larger Portion	(4) C Area in Smaller Portion	(5) y Ordinate at z
1.87	.4693	.9693	.0307	.0694
1.88	.4699	.9699	.0301	.0681
1.89	.4706	.9706	.0294	.0669
1.90	.4713	.9713	.0287	.0656
1.91	.4719	.9719	.0281	.0644
1.92	.4726	.9726	.0274	.0632
1.93	.4732	.9732	.0268	.0620
1.94	.4738	.9738	.0262	.0608
1.95	.4744	.9744	.0256	.0596
1.96	.4750	.9750	.0250	.0584
1.97	.4756	.9756	.0244	.0573
1.98	.4761	.9761	.0239	.0562
1.99	.4767	.9767	.0233	.0551
2.00	.4772	.9772	.0228	.0540
2.01	.4778	.9778	.0222	.0529
2.02	.4783	.9783	.0217	.0519
2.03	.4788	.9788	.0212	.0508
2.04	.4793	.9793	.0207	.0498
2.05	.4798	.9798	.0202	.0488
2.06	.4803	.9803	.0197	.0478
2.07	.4808	.9808	.0192	.0468
2.08	.4812	.9812	.0188	.0459
2.09	.4817	.9817	.0183	.0449
2.10	.4821	.9821	.0179	.0440
2.11	.4826	.9826	.0174	.0431
2.12	.4830	.9830	.0170	.0422
2.13	.4834	.9834	.0166	.0413
2.14	.4838	.9838	.0162	.0404
2.15	.4842	.9842	.0158	.0396
2.16	.4846	.9846	.0154	.0387
2.17	.4850	.9850	.0150	.0379
2.18	.4854	.9854	.0146	.0371
2.19	.4857	.9857	.0143	.0363
2.20	.4861	.9861	.0139	.0355
2.21	.4864	.9864	.0136	.0347
2.22	.4868	.9868	.0132	.0339
2.23	.4871	.9871	.0129	.0332
2.24	.4875	.9875	.0125	.0325
2.25	.4878	.9878	.0122	.0317
2.26	.4881	.9881	.0119	.0310
2.27	.4884	.9884	.0116	.0303
2.28	.4887	.9887	.0113	.0297
2.29	.4890	.9890	.0110	.0290

2.30	.4893	.9893	.0107	.0283
2.31	.4896	.9896	.0104	.0277
2.32	.4898	.9898	.0102	.0270
2.33	.4901	.9901	.0099	.0264
2.34	.4904	.9904	.0096	.0258
2.35	.4906	.9906	.0094	.0252
2.36	.4909	.9909	.0091	.0246
2.37	.4911	.9911	.0089	.0241
2.38	.4913	.9913	.0087	.0235
2.39	.4916	.9916	.0084	.0229
2.40	.4918	.9918	.0082	.0224
2.41	.4920	.9920	.0080	.0219
2.42	.4922	.9922	.0078	.0213
2.43	.4925	.9925	.0075	.0208
2.44	.4927	.9927	.0073	.0203
2.45	.4929	.9929	.0071	.0198
2.46	.4931	.9931	.0069	.0194
2.47	.4932	.9932	.0068	.0189
2.48	.4934	.9934	.0066	.0184
2.49	.4936	.9936	.0064	.0180
2.50	.4938	.9938	.0062	.0175
2.51	.4940	.9940	.0060	.0171
2.52	.4941	.9941	.0059	.0167
2.53	.4943	.9943	.0057	.0163
2.54	.4945	.9945	.0055	.0158
2.55	.4946	.9946	.0054	.0154
2.56	.4948	.9948	.0052	.0151
2.57	.4949	.9949	.0051	.0147
2.58	.4951	.9951	.0049	.0143
2.59	.4952	.9952	.0048	.0139
2.60	.4953	.9953	.0047	.0136
2.61	.4955	.9955	.0045	.0132
2.62	.4956	.9956	.0044	.0129
2.63	.4957	.9957	.0043	.0126
2.64	.4959	.9959	.0041	.0122
2.65	.4960	.9960	.0040	.0119
2.66	.4961	.9961	.0039	.0116
2.67	.4962	.9962	.0038	.0113
2.68	.4963	.9963	.0037	.0110
2.69	.4964	.9964	.0036	.0107
2.70	.4965	.9965	.0035	.0104
2.71	.4966	.9966	.0034	.0101
2.72	.4967	.9967	.0033	.0099
2.73	.4968	.9968	.0032	.0096
2.74	.4969	.9969	.0031	.0093
2.75	.4970	.9970	.0030	.0091
2.76	.4971	.9971	.0029	.0088
2.77	.4972	.9972	.0028	.0086
2.78	.4973	.9973	.0027	.0084
2.79	.4974	.9974	.0026	.0081
2.80	.4974	.9974	.0026	.0079
2.81	.4975	.9975	.0025	.0077
2.82	.4976	.9976	.0024	.0075
2.83	.4977	.9977	.0023	.0073
2.84	.4977	.9977	.0023	.0071

Table F (cont.)

(1) z Standard Score $\left(\frac{x}{\sigma}\right)$	(2) A Area from Mean to $\frac{x}{\sigma}$	(3) B Area in Larger Portion	(4) C Area in Smaller Portion	(5) y Ordinate at $\frac{x}{\sigma}$
2.85	.4978	.9978	.0022	.0069
2.86	.4979	.9979	.0021	.0067
2.87	.4979	.9979	.0021	.0065
2.88	.4980	.9980	.0020	.0063
2.89	.4981	.9981	.0019	.0061
2.90	.4981	.9981	.0019	.0060
2.91	.4982	.9982	.0018	.0058
2.92	.4982	.9982	.0018	.0056
2.93	.4983	.9983	.0017	.0055
2.94	.4984	.9984	.0016	.0053
2.95	.4984	.9984	.0016	.0051
2.96	.4985	.9985	.0015	.0050
2.97	.4985	.9985	.0015	.0048
2.98	.4986	.9986	.0014	.0047
2.99	.4986	.9986	.0014	.0046
3.00	.4987	.9987	.0013	.0044
3.01	.4987	.9987	.0013	.0043
3.02	.4987	.9987	.0013	.0042
3.03	.4988	.9988	.0012	.0040
3.04	.4988	.9988	.0012	.0039
3.05	.4989	.9989	.0011	.0038
3.06	.4989	.9989	.0011	.0037
3.07	.4989	.9989	.0011	.0036
3.08	.4990	.9990	.0010	.0035
3.09	.4990	.9990	.0010	.0034
3.10	.4990	.9990	.0010	.0033
3.11	.4991	.9991	.0009	.0032
3.12	.4991	.9991	.0009	.0031
3.13	.4991	.9991	.0009	.0030
3.14	.4992	.9992	.0008	.0029
3.15	.4992	.9992	.0008	.0028
3.16	.4992	.9992	.0008	.0027
3.17	.4992	.9992	.0008	.0026
3.18	.4993	.9993	.0007	.0025
3.19	.4993	.9993	.0007	.0025
3.20	.4993	.9993	.0007	.0024
3.21	.4993	.9993	.0007	.0023
3.22	.4994	.9994	.0006	.0022
3.23	.4994	.9994	.0006	.0022
3.24	.4994	.9994	.0006	.0021
3.30	.4995	.9995	.0005	.0017
3.40	.4997	.9997	.0003	.0012
3.50	.4998	.9998	.0002	.0009
3.60	.4998	.9998	.0002	.0006
3.70	.4999	.9999	.0001	.0004

Bibliography

For those who might like to forge ahead on their own, there are now many good, rather clear texts. The following brief personal list omits many good books and is in the spirit of some suggestions the author would give if someone were to ask, "Where would be a good place to look for more detail on _____?".

Elementary texts that cover about the same content as this book but go into greater depth and provide some extensions are:

Downie, N. M., and R. W. Heath. *Basic Statistical Methods*, 4th ed. New York: Harper & Row, 1974. A popular introductory text.

Roscoe, J. T. *Fundamental Research Statistics*. New York: Holt, Rinehart & Winston, 1969. Particularly strong at illustrating relationships between different techniques.

Advanced texts that could now be understood by the reader but that go well beyond the scope of this book are:

Guilford, J. P., and Benjamin Fruchter. *Fundamental Statistics in Psychology and Education*, 5th ed. New York: McGraw-Hill, 1973.

Hays, William L. *Statistics for the Social Sciences*, 2nd ed. New York: Holt, Rinehart & Winston, 1973.

Evaluative Data

The quality of programed instruction depends strongly on feedback cycles; the author creates some material and tries it out on students. The tryout reveals sections that teach particularly efficiently, sections so repetitive that they are dull, and sections so dense that they baffle. The cycle closes as the author, using the results of criterion testing—filled-in programs, interviews with students, and questions raised during class—prepares a new version which hopefully corrects any earlier problems but almost invariably introduces some new (usually fewer) difficulties, and distributes this new material for yet another try.

Repeated cycling with diverse students is an ongoing process resulting in material that, while always short of perfection, gets constantly better.

Samples used in feedback cycles up to now include:

Level	Course	School
Freshman and Sophomore	Introductory Psychology	Ithaca College Ithaca, N. Y.
Undergraduate	Educational Psychology	Brigham Young University Provo, Utah
Undergraduate	Programmed Learning, Educational Psychology	University of Toledo Toledo, Ohio
Undergraduate	Introductory Psychology	Cornell University Ithaca, N. Y.
Graduate	Guidance	
Undergraduate and Graduate	Introductory Statistics and Educational Measurement	Jackson State College Jackson, Miss.
Undergraduate	Introductory Statistics (for nine successive years), Experimental Psychology	State University College at Oneonta Oneonta, N. Y.
High School	Special	Four Upper N. Y. State Counties
Seventh Grade	Special	Bugbee School Oneonta, N. Y.

In addition to these specific studies, many helpful suggestions and criticisms have been received from students and teachers during the nine years since the first edition of *Statistical Concepts: A Basic Program* was published. This frequently detailed and insightful informal feedback has led to several significant improvements in this second edition.

92 93 94 95 96 30 29 28 27 26 25 24